DYNAMICS STUDY PACK

FREE-BODY DIAGRAM WORKBOOK & CHAPTER REVIEWS

PETER SCHIAVONE
UNIVERSITY OF ALBERTA

Engineering Mechanics
DYNAMICS

Bedford • Fowler

FOURTH EDITION

Upper Saddle River, NJ 07458

Executive Editor: *Eric Svendsen*
Associate Editor: *Dee Bernhard*
Executive Managing Editor: *Vince O'Brien*
Managing Editor: *David A. George*
Production Editor: *Craig Little*
Supplement Cover Manager: *Paul Gourhan*
Supplement Cover Designer: *Joanne Alexandris*
Manufacturing Buyer: *Ilene Kahn*

Pearson Prentice Hall
Pearson Education, Inc.
Upper Saddle River, NJ 07458

Printed in the United States of America

10 9 8 7 6 5 4 3 2 1

ISBN 0-13-150290-5

Pearson Education Ltd., *London*
Pearson Education Australia Pty. Ltd., *Sydney*
Pearson Education Singapore, Pte. Ltd.
Pearson Education North Asia Ltd., *Hong Kong*
Pearson Education Canada, Inc., *Toronto*
Pearson Educación de Mexico, S.A. de C.V.
Pearson Education—Japan, *Tokyo*
Pearson Education Malaysia, Pte. Ltd.
Pearson Education, Inc., *Upper Saddle River, New Jersey*

Contents

Preface

This supplement is divided into two parts. Part I is a workbook which explains how to draw and use free-body diagrams when solving problems in Dynamics. Part II provides a section-by-section, chapter-by-chapter summary of the key concepts, principles and equations from A. Bedford & W. Fowler's text: *Engineering Mechanics DYNAMICS (4th Edition).*

Part I. Free-Body Diagram Workbook

A thorough understanding of how to draw and use a free-body diagram is absolutely essential when solving problems in mechanics.

This workbook consists mainly of a collection of problems intended to give the student practice in drawing and using free-body diagrams when solving problems in *Dynamics.*

All the problems are presented as *tutorial* problems with the solution only partially complete. The student is then expected to complete the solution by 'filling in the blanks' in the spaces provided. This gives the student the opportunity to *build free-body diagrams in stages* and extract the relevant information from them when formulating equilibrium equations. Earlier problems provide students with partially drawn free-body diagrams and lots of hints to complete the solution. Later problems are more advanced and are designed to challenge the student more. The complete solution to each problem can be found at the back of the page. The problems are chosen from two-dimensional theories of particle and rigid-body dynamics. Once the ideas and concepts developed in these problems have been understood and practiced, the student will find that they can be extended in a relatively straightforward manner to accommodate the corresponding three-dimensional theories.

The book begins with a brief primer on free-body diagrams: where they fit into the general procedure of solving problems in mechanics and why they are so important. Next follows a few examples to illustrate ideas and then the workbook problems.

For best results, the student should read the primer and then, beginning with the simpler problems, try to complete and understand the solution to each of the subsequent problems. The student should avoid the temptation to immediately look at the completed solution over the page. This solution should be accessed only as a last resort (after the student has struggled to the point of giving up), or to check the student's own solution after the fact. The idea behind this is very simple:

We learn most when we actually ***do*** *the thing we are trying to learn.*

In other words, reading through someone else's solution is not the same as actually working through the problem. In the former, the student gains *information*, in the latter the student gains *knowledge*. For example, how many people learn to swim or drive a car by reading an instruction manual?

Consequently, since this book is based on ***doing***, the student who persistently solves the problems in this book will ultimately gain a thorough, usable knowledge of how to draw and use free-body diagrams.

Part II. Chapter-by-Chapter Summaries

This part of the supplement provides a section-by-section, chapter-by-chapter summary of the key concepts, principles and equations from A. Bedford & W. Fowler's text: *Engineering Mechanics DYNAMICS (4th Edition).* We follow the same section and chapter order as that used in the *text* and summarize important concepts from each section in easy-to-understand language. We end each chapter summary with a simple set of review questions designed to see if the student has understood the key concepts and chapter objectives.

This section of the supplement will be useful both as a quick reference guide for important concepts and equations when solving problems in, for example, homework assignments or laboratories and also as a handy review when preparing for any quiz, test or examination.

P. Schiavone

PART I

FREE-BODY DIAGRAM WORK BOOK

1

Basic Concepts in Dynamics

Engineering mechanics is divided into two areas: statics and dynamics. *Statics* deals with the equilibrium of bodies, that is, those that are either at rest (if originally at rest) or move with constant velocity (if originally in motion). *Dynamics* is concerned with the accelerated motion of bodies. The study of dynamics is itself divided into two parts: *kinematics*, which treats only the geometric aspects of motion and *kinetics* which is concerned with the analysis of forces causing the motion. Free-body diagrams play a significant role in solving problems in *kinetics*.

In mechanics, real bodies (e.g. planets, cars, planes, tables, crates, etc) are represented or *modeled* using certain idealizations which simplify application of the relevant theory. In this book we refer to only two such models:

- **Particle or Point in Space.** *A particle* has mass but no size/shape. When an object's size/shape can be neglected so that only its mass is relevant to the description of its motion, the object can be modeled as a particle. This is the same thing as saying that the motion of the object can be modeled as the motion of a *point in space* (the point itself representing the center of mass of the moving object). For example, the size of an aircraft is insignificant when compared to the size of the earth and therefore the aircraft can be modeled as a particle (or point in space) when studying its three dimensional motion in space.
- **Rigid Body.** A *rigid body* represents the next level of modeling sophistication after the particle. That is, a rigid body is a collection of particles (which therefore has mass) which has a significant size/shape but this size/shape cannot change. In other words, when an object is modeled as a rigid body, we assume that any deformations (changes in shape) are relatively small and can be neglected. Although any object does deform as it moves, if its deformation is small, *you can approximate its motion by modeling it as a rigid body*. For example, the actual deformations occurring in most structures and machines are relatively small so that the rigid body assumption is suitable in these cases.

1.1 Equations of Motion

1.1.1 Equation of Motion for a Particle

When a system of forces acts on a particle, the equation of motion may be written in the form

$$\Sigma \mathbf{F} = m\mathbf{a} \tag{1.1}$$

where $\Sigma \mathbf{F}$ is the vector sum of all the external forces acting on the particle and m and $\mathbf{a}$ are, respectively, the mass and acceleration of the particle.

Successful application of the equation of motion (1.1) requires a complete specification of all the known and unknown external forces ($\Sigma\mathbf{F}$) that act on the object. The best way to account for these is to draw the object's *free-body diagram*: a sketch of the object *freed* from its surroundings showing *all* the (external) forces that *act* on it. In dynamics problems, since the resultant of these external forces produces the vector $m\mathbf{a}$, in addition to the free-body diagram, a *kinetic diagram* is often used to represent graphically the magnitude and direction of the vector $m\mathbf{a}$. In other words, the equation (1.1) can be represented graphically as:

Free-body Diagram = Kinetic Diagram

Of course, whenever the equation of motion (1.1) is applied, it is required that measurements of the acceleration be made from a *Newtonian* or inertial frame of reference. *Such a coordinate system does not rotate and is either fixed or translates in a given direction with a constant velocity (zero acceleration).* This definition ensures that the particle's acceleration measured by observers in two different inertial frames of reference will always be the *same*.

1.1.2 Equation of Motion for a System of Particles

The equation of motion (1.1) can be extended to include a *system of particles* isolated within an enclosed region in space:

$$\Sigma\mathbf{F} = m\mathbf{a}_G \tag{1.2}$$

This equation states that the sum of external forces ($\Sigma\mathbf{F}$) acting on the system of particles is equal to the total mass m of the particles multiplied by the acceleration $\mathbf{a}_G$ of its mass center G. Since, in reality, all particles must have a finite size to possess mass, equation (1.2) justifies application of the equation of motion to a *body* that is represented as a single particle.

1.1.3 Equations of Motion for a Rigid Body

Since rigid bodies, by definition, have a definite size/shape, their motion is governed by *both* translational and rotational quantities. The translational equation of motion for (the mass center of) a rigid body is basically equation (1.2). That is,

$$\Sigma\mathbf{F} = m\mathbf{a}_G \tag{1.2}$$

In this case, the equation (1.2) states that the sum of all the external forces acting on the body is equal to the body's mass m multiplied by the acceleration $\mathbf{a}_G$ of its mass center G.

The rotational equation of motion for a rigid body is given by

$$\Sigma\mathbf{M}_G = I_G\boldsymbol{\alpha} \tag{1.3}$$

which states that the sum of the applied couple moments and moments of all the external forces computed about a body's mass center $G(\Sigma\mathbf{M}_G)$ is equal to the product of the moment of inertia of the body about an axis passing through $G(I_G)$ and the body's angular acceleration $\boldsymbol{\alpha}$.

Alternatively, equation (1.3) can be re-written in more general form as:

$$\Sigma\mathbf{M}_P = I_G\boldsymbol{\alpha} + \Sigma(\mathbf{M}_a)_P \tag{1.4}$$

Here, $\Sigma\mathbf{M}_P$ represents the sum of the applied couple moments and the external moments taken about a general point $P(\not\equiv G)$ and $\Sigma(\mathbf{M}_a)_P$ represents the moments generated by the components of the vector $m\mathbf{a}_G$ about the point P.

When applying the equations of motion (1.2)–(1.4), one should always draw a *free-body diagram* in order to account for the terms involved in ($\Sigma\mathbf{F}$), ($\Sigma\mathbf{M}_G$), or ($\Sigma\mathbf{M}_P$). The *kinetic diagram* is also useful in that it accounts graphically for the acceleration components $m\mathbf{a}_G$ and the term $I_G\boldsymbol{\alpha}$, and it is especially convenient when used to determine the moment terms $\Sigma(\mathbf{M}_a)_P$ generated by the components of the vector $m\mathbf{a}_G$ about the point P (when using (1.4)).

2

Free-Body Diagrams: the Basics

2.1 Free-Body Diagram: Particle

The equation of motion (1.1) is used to analyze the motion of *an object* (modeled as a particle) when subjected to an unbalanced force system. The first step in this analysis is to draw the *free-body* diagram to identify the external forces ($\Sigma\mathbf{F}$) acting on the object. The free-body diagram is simply a sketch of the object *freed* from its surroundings showing *all* the (external) forces that *act* on the object. The diagram focuses your attention on the object of interest and helps you identify all the external forces ($\Sigma\mathbf{F}$) acting. Once the free-body diagram is drawn, it may be helpful to draw the corresponding *kinetic diagram*. This diagram accounts graphically for the acceleration components (components of the vector a) on the object. Taken together, these diagrams provide (in graphical form) all the information that is needed to write down the equation of motion (1.1).

EXAMPLE 2.1

The crate A shown in Figure 1, is released from rest. Its mass is $m_A = 50$ kg and the coefficient of kinetic friction between the crate and the inclined surface is $\mu_k = 0.15$. Draw the free-body and kinetic diagrams of crate A as it slides down the plane.

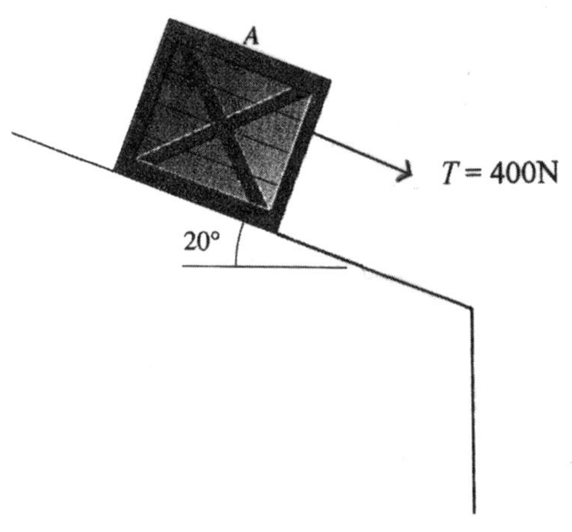

Figure 1

Solution

The free-body diagram of crate A is shown in Figure 2(a). Notice that once the crate is *separated* or *freed* from the system (= crate + plane), forces which were previously internal to the system become external to the crate. For example, in Figure 2(a), such a force is the force of friction *acting on the crate.* The kinetic diagram is shown in Figure 2(b). In this case, the diagram shows the components of the crate's acceleration vector **a**.

Next, we present a formal procedure for drawing free-body diagrams for a particle or system of particles. ◀

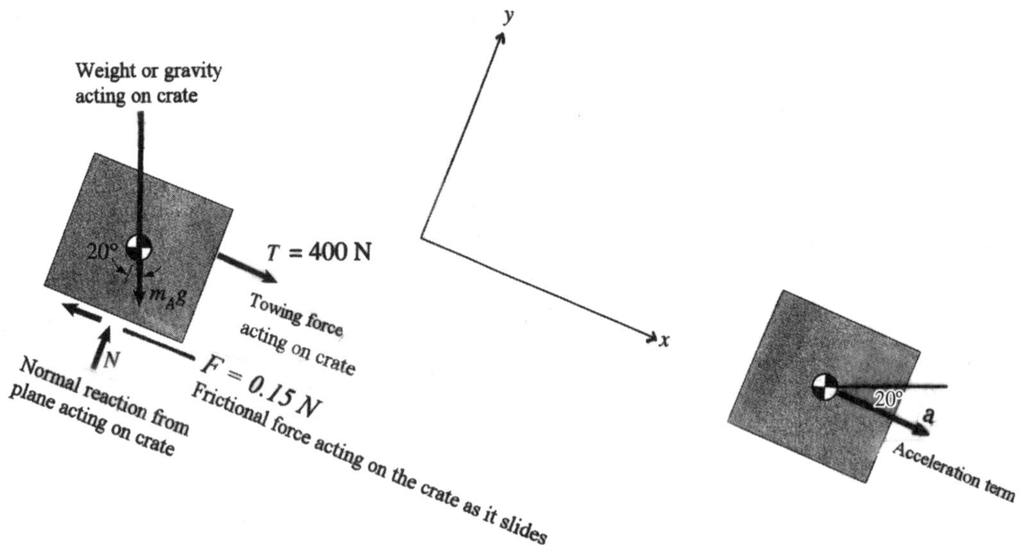

(a) Free-body diagram of crate A.

(b) The crate's acceleration parallel to the plane.

Figure 2

2.1.1 Procedure for Drawing a Free-Body Diagram: Particle.

1. *Select* the inertial coordinate system. Most often, rectangular or x, y-coordinates are chosen to analyze problems for which the particle has *rectilinear motion.* If this occurs, one of the axes should extend in the direction of motion.
2. *Identify the object you wish to isolate* from the system. This choice is often dictated by the particular forces of interest.
3. *Draw the outlined shape of the isolated object.* Imagine the object to be isolated or cut free from the system of which it is a part.
4. *Show all external forces acting on the isolated object.* Indicate on this sketch all the external forces that act on the object. These forces can be *active forces*, which tend to set the object in motion, or they can be *reactive forces* which are the result of the constraints or supports that prevent motion. This stage is crucial: it may help to trace around the object's boundary, carefully noting each external force acting on it. Don't forget to include the weight of the object (unless it is being intentionally neglected).
5. *Identify and label each external force acting on the (isolated) object.* The forces that are known should be labeled with their known magnitudes and directions. Use letters to represent the magnitudes and arrows to represent the directions of forces that are unknown.
6. *The direction of a force having an unknown magnitude can be assumed.*
7. *The direction and sense* of the particle's acceleration **a** should also be established. If the sense of its components is unknown, assume they are in the same direction as the positive inertial coordinate axes. The acceleration may be sketched on the x, y-coordinate system or it may be represented on the *kinetic diagram.*

2.1.2 *Using the Free-Body Diagram: Equations of Motion*

The equations of motion (1.1) or (1.2) are used to solve problems which require a relationship between the forces acting on a particle and the accelerated motion they cause. Whenever (1.1) or (1.2) is applied, the unknown force and acceleration components should be identified and an equivalent number of equations should be written. If further equations are required for the solution, kinematics may be considered.

The*free-body diagram* is used to identify the unknown force and the *kinetic diagram* the unknown acceleration components acting on the particle. The subsequent procedure for solving problems once the free-body (and, if necessary, the kinetic) diagram for the particle is established, is therefore as follows:

1. If the forces can be resolved directly from the free-body diagram, apply the equations of motion in their scalar component form. For example:
$$\Sigma F_x = ma_x \text{ and } \Sigma F_y = ma_y \tag{2.1}$$
2. Components are positive if they are directed along a positive axis and negative if they are directed along a negative axis.
3. If the particle contacts a rough surface, it may be necessary to use the frictional equation, which relates the coefficient of kinetic friction to the magnitudes of the frictional and normal forces acting at the surfaces of contact. Remember that the frictional force always acts on the free-body diagram such that it opposes the motion of the particle *relative to the surface it contacts.*
4. If the solution yields a negative result, this indicates the sense of the force is the reverse of that shown/assumed on the free-body diagram.

EXAMPLE 2.2

In Example 2.1, the diagrams established in Figure 2 give us a "pictorial representation" of all the information we need to apply the equations of motion (2.1) to find the unknown force **N** and the acceleration **a**. In fact, taking the positive x-direction to be parallel to the plane ($\searrow +$) and the positive y-direction to be perpendicular to the plane ($\nearrow +$),the equations of motion (2.1) when applied to crate A (regarded as a particle—since its shape is not important in the motion under consideration) are:

For Crate A:
$$\searrow +\Sigma F_x = ma_x: \quad 400 + 50\,\text{g}\sin 20^\circ - F = 50a_x$$
$$\nearrow +\Sigma F_y = ma_y: \quad N - 50\,\text{g}\cos 20^\circ = 0$$

Two equations, 3 unknowns: use the frictional equation to relate F to N and obtain a third equation:

$$\text{Frictional Equation (block is sliding):} \quad F = 0.15\,\text{N}.$$

Solving these three equations yields

$$N = 460.92\,\text{N}, \quad a_x = 9.97\,\text{m/s}^2 \quad \text{Ans.}$$

The directions of each of the vectors N and **a** is shown in the free-body diagram above (Figure 2). ◀

2.2 Free-Body Diagram: Rigid Body

The equations of motion (1.2) and (1.3) (or (1.4)) are used to determine unknown forces, moments and acceleration components acting *on an object* (modeled as a rigid body) subjected to an unbalanced system of forces and moments. The first step in doing this is again to draw the *free-body* diagram of the object to identify *all* of the external forces and moments acting on it. The procedure for drawing a free-body diagram in this case is much the same as that for a particle with the main difference being that now, because the object has "size/shape", it can support also external couple moments and moments of external forces.

2.2.1 Procedure for Drawing a Free-Body Diagram: Rigid Body

1. Select the inertial x, y or n, t coordinate system. This will depend on whether the body is in rectilinear or curvilinear motion.
2. Imagine the body to be isolated or 'cut free' from its constraints and connections and sketch its outlined shape.
3. Identify all the external forces and couple moments that act on the body. Those generally encountered are:
   (a) Applied loadings
   (b) Reactions occurring at the supports or at points of contact with other bodies.
   (c) The weight of the body (applied at the body's center of gravity G)
   (d) Frictional forces
4. The forces and couple moments that are known should be labeled with their proper magnitudes and directions. Letters are used to represent the magnitudes and direction angles of forces and couple moments that are *unknown*. Indicate the dimensions of the body necessary for computing the moments of external forces. In particular, if a force or couple moment has a known line of action but unknown magnitude, the arrowhead which defines the sense of the vector can be assumed. The correctness of the assumed sense will become apparent after solving the equations of motion for the unknown magnitude. By definition, the magnitude of a vector is *always positive*, so that if the solution yields a *negative* scalar, the minus sign indicates that the vector's sense is *opposite* to that which was originally assumed.
5. *The direction and sense* of the acceleration of the body's mass center $\mathbf{a}_G$ should also be established. If the sense of its components is unknown, assume they are in the same direction as the positive inertial coordinate axes. The acceleration may be sketched on the x, y-coordinate system or it may be represented as the $\mathbf{a}_G$ vector on the *kinetic diagram*. This will also be helpful for "visualizing" the terms needed in the moment sum $\Sigma(\boldsymbol{M}_a)_P$ of equation (1.4), since the kinetic diagram accounts graphically for the components $(\mathbf{a}_G)_x$, $(\mathbf{a}_G)_x$, or $(\mathbf{a}_G)_t$, $(\mathbf{a}_G)_n$.

Important Points

- Internal forces are never shown on the free-body diagram since they occur in equal but opposite collinear pairs and therefore cancel each other out.
- The weight of a body is an external force and its effect is shown as a single resultant force acting through the body's center of gravity G.
- *Couple moments* can be placed anywhere on the free-body diagram since they are *free vectors*. Forces can act at any point along their lines of action since they are *sliding vectors*.

EXAMPLE 2.3

Draw the free-body and kinetic diagrams for the 60-kg crate. A horizontal force of 700 N is applied to the crate as shown and the coefficient of kinetic friction between the crate and the ground is $\mu_k = 0.3$.

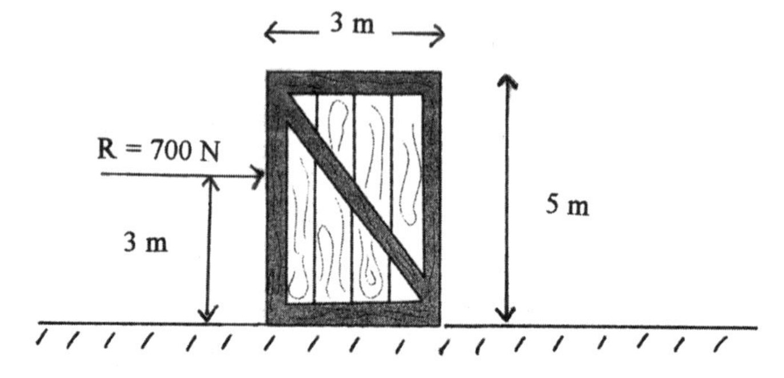

Figure 3

Solution

Here, since the force **R** can cause the crate to either slide or to tip over, we model the crate as a rigid body. This model allows us to account for the effects of moments arising from **R** and any other external forces. We begin by assuming that the crate slides so that the frictional equation yields $F = \mu_k, N_C = 0.3N_C$. Also, the normal force $\mathbf{N}_C$ acts at O, a distance s (where $0 < s \leq 1.5$ m) from the crate's center line. Note that the line of action of $\mathbf{N}_C$ does not necessarily pass through the mass center G ($s = 0$), since $\mathbf{N}_C$ must counteract the tendency for tipping caused by **R**.

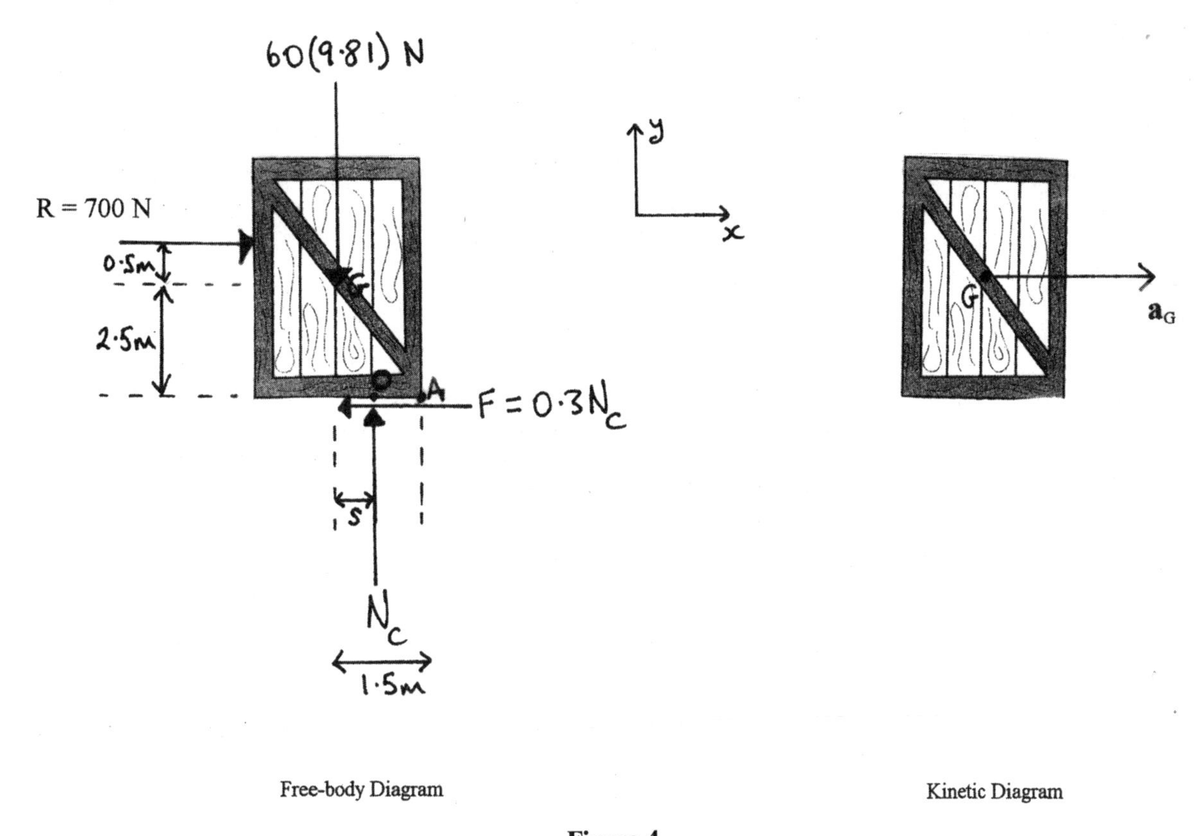

Free-body Diagram — Kinetic Diagram

Figure 4

(Note that had we assumed that the crate tips, then the normal force $\mathbf{N}_C$ would have been assumed to act at the corner point A and the frictional equation would take the form $F \leq 0.3N_C$). ◀

2.2.2 Using the Free-Body Diagram: Equations of Motion

The procedure for solving kinetic problems for a rigid body once the free-body diagram is established, is as follows:

- Apply the three equations of motion (1.2)–(1.3). To simplify the analysis, the moment equation (1.3) may be replaced by the more general equation (1.4) where the point P is usually located at the intersection of the lines of action of as many unknown forces as possible.
- If the body contacts a rough surface, it may be necessary to use the frictional equation, which relates the coefficient of kinetic friction to the magnitudes of the frictional and normal forces acting at the surfaces of contact. Remember that the frictional force always acts on the free-body diagram such that it *opposes the motion of the body relative to the surface it contacts.*
- Use kinematics if the velocity and position of the body are to be determined.

EXAMPLE 2.4

Find the acceleration of the crate in Example 2.3.

Solution

Using the free-body diagram in Figure 4, the equations of motion are (taking counterclockwise as positive):

$$\rightarrow +\Sigma F_x = m(a_G)_x : \quad 700\text{ N} - 0.3N_C = (60\text{ kg})(a_G)_x$$

$$\uparrow +\Sigma F_y = m(a_G)_y : \quad N_C - 589.2\text{ N} = 0$$

$$\Sigma M_G = I_G\alpha : \quad -700\text{ N}(0.5\text{ m}) + N_C(s) - 0.3N_C(2.5\text{ m}) = 0$$

Solving, we obtain $N_C = 589.2\text{ N} \uparrow, s = 1.34\text{ m}, a_G = 8.7\text{ m/s}^2 \rightarrow$ *Ans.*

Since $s = 1.34\text{ m} < 1.5\text{ m}$, indeed the crate slides as originally assumed (otherwise the problem would have to be reworked with the assumption that tipping occurred). ◀

3

Problems

3.1 Free-Body Diagrams in Particle Kinetics

Problem 3.1

The 2–kg collar A is initially at rest on the smooth horizontal bar. At $t = 0$ (where t is time),the collar is subjected to a constant horizontal force with magnitude $F = 4$ N as shown. Draw the free-body and kinetic diagrams for the collar.

Solution

1. The size/shape of the collar does not affect the (rectilinear) motion under consideration. Consequently, we assume that the collar has *negligible size* so that it can be modelled as a particle.
2. Imagine the collar to be separated or detached from the system (collar + bar).
3. The (detached) collar is subjected to three *external* forces. They are caused by:

   i. **ii.**

   iii.
4. Draw the free-body diagram of the (detached) collar showing all these forces labeled with their magnitudes and directions. Include any other information e.g. angles, lengths etc which may help when formulating the equations of motion.
5. The acceleration of the collar is along the bar. Show this on a kinetic diagram or on the inertial coordinate system chosen in the free-body diagram.

Problem 3.1

The 2−kg collar A is initially at rest on the smooth horizontal bar. At $t = 0$ (where t is time), the collar is subjected to a constant horizontal force with magnitude $F = 4$ N as shown. Draw the free-body and kinetic diagrams for the collar.

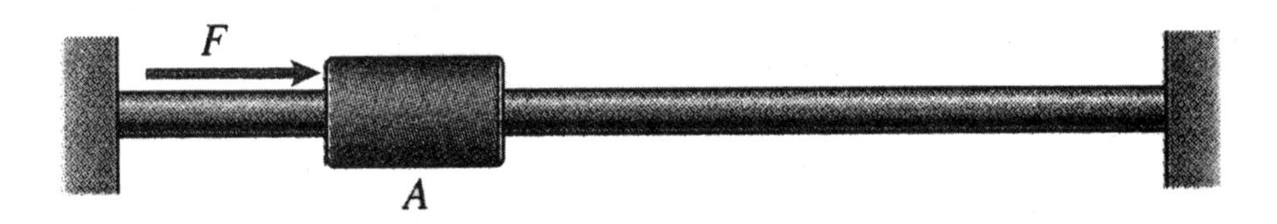

Solution

1. The size/shape of the collar does not affect the (rectilinear) motion under consideration. Consequently, we assume that the collar has *negligible size* so that it can be modelled as a particle.
2. Imagine the collar to be separated or detached from the system (collar + bar).
3. The (detached) collar is subjected to three *external* forces. They are caused by:
   i. **Collar's Weight**
   ii. **Bar's Reaction to Collar**
   iii. **Force with Magnitude** $F = 4$ N
4. Draw the free-body diagram of the (detached) collar showing all these forces labeled with their magnitudes and directions. Include any other information e.g. angles, lengths etc which may help when formulating the equations of motion.
5. The acceleration of the collar is along the bar. Show this on a kinetic diagram or on the inertial coordinate system chosen in the free-body diagram.

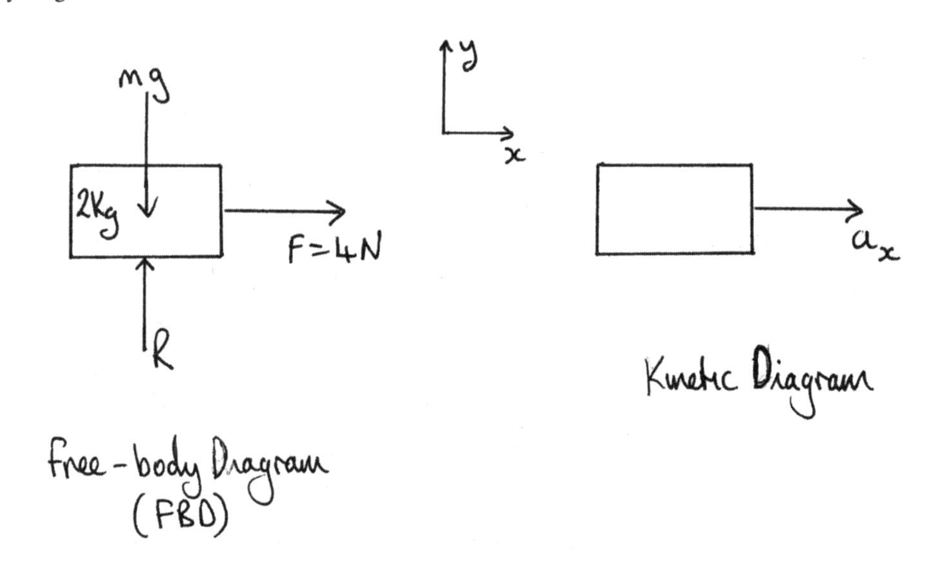

Problem 3.2

Suppose in Problem 3.1, the bar is no longer smooth and the coefficient of kinetic friction between the collar and the bar is $\mu_k = 0.1$. Draw the free-body and kinetic diagrams for the collar and use them to determine how fast the collar is moving and how far the collar has travelled after one second.

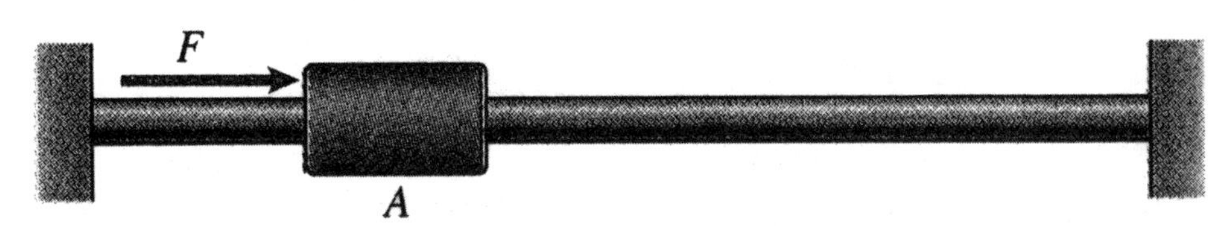

Solution

1. The size/shape of the collar does not affect the (rectilinear) motion under consideration. Consequently, we assume that the collar has *negligible size* so that it can be modelled as a particle.
2. Imagine the collar to be separated or detached from the system (collar + bar).
3. The (detached) collar is subjected to four *external* forces

   i. **ii.**

   iii. **iv.**

4. Draw the free-body diagram of the (detached) collar showing all these forces labeled with their magnitudes and directions. Include any other information e.g. angles, lengths etc which may help when formulating the equations of motion.
5. Establish an appropriate $xy-$axes system and draw the corresponding kinetic diagram.

6. Using the $xy-$axes system on the free-body diagram, write down the equations of motion for the collar in the $x-$ and $y-$directions:

   $+\!\rightarrow \sum F_x = ma_x$:

   $+\!\uparrow \sum F_y = ma_y$:

7. Solve for the acceleration of the collar.
8. Integrate the acceleration of the collar to obtain the desired speed and distance:

Problem 3.2

Suppose in Problem 3.1, the bar is no longer smooth and the coefficient of kinetic friction between the collar and the bar is $\mu_k = 0.1$. Draw the free-body and kinetic diagrams for the collar and use them to determine how fast the collar is moving and how far the collar has travelled after one second.

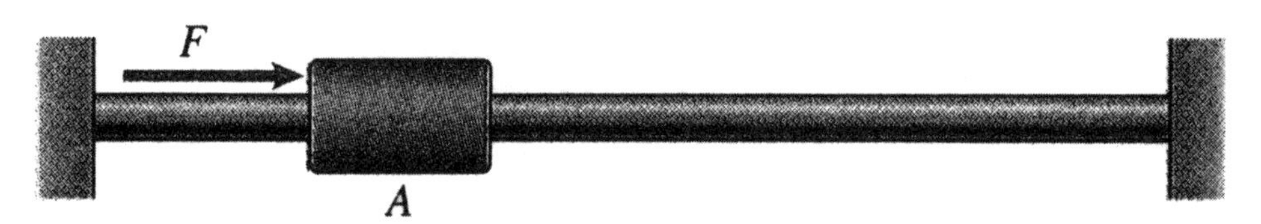

Solution

1. The size/shape of the collar does not affect the (rectilinear) motion under consideration. Consequently, we assume that the collar has *negligible size* so that it can be modelled as a particle.
2. Imagine the collar to be separated or detached from the system (collar + bar).
3. The (detached) collar is subjected to four *external* forces. They are caused by:

   i. Collar's Weight **ii. Bar's Reaction to Collar**

   iii. Force with Magnitude $F = 4$ N **iv. Friction Force on Collar**

4. Draw the free-body diagram of the (detached) collar showing all these forces labeled with their magnitudes and directions. Include any other information e.g. angles, lengths etc which may help when formulating the equations of motion.
5. Establish an appropriate $xy-$axes system and draw the corresponding kinetic diagram.

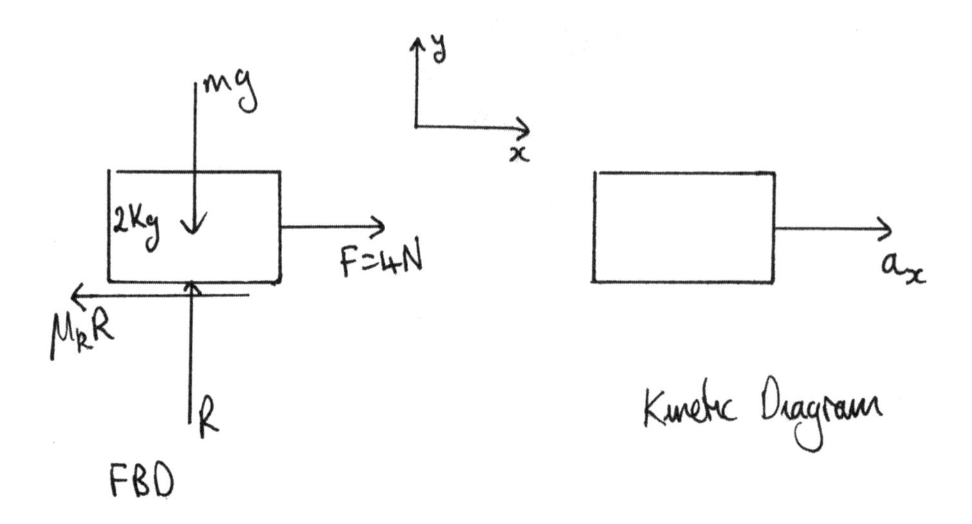

6. Using the $xy-$axes system on the free-body diagram, write down the equations of motion for the collar in the $x-$ and $y-$ directions:

   $\rightarrow + \sum F_x = ma_x : 4 - \mu_k R = 2a_x \Leftrightarrow 4 - (0.1)\, R = 2a_x$

   $+\uparrow \ \sum F_y = ma_y : R - mg = 0 \Leftrightarrow R - 2g = 0$
7. Solve for the acceleration of the collar. $a_x = 1.02 \text{ m/s}^2$, $R = 19.62$ N
8. Integrate the acceleration of the collar to obtain the desired speed and distance:

   $v_x = a_x t$, $x = a_x \frac{t^2}{2}$. Now let $t = 1$ s to obtain $v_x = 1.02$ m/s, $x = 0.51$ m.

Problem 3.3

The 20–lb collar A is initially at rest on the smooth bar. At $t = 0$ (where t is time), it is subjected to a constant force with magnitude $F = 10$ lb as shown. Draw the free-body and kinetic diagrams for the collar.

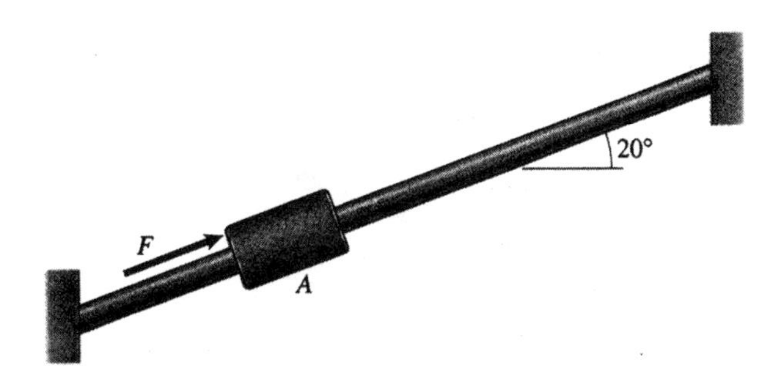

Solution

1. The size/shape of the collar does not affect the (rectilinear) motion under consideration. Consequently, we assume that the collar has *negligible size* so that it can be modelled as a particle.
2. Imagine the collar to be separated or detached from the system (collar + bar).
3. The (detached) collar is subjected to three *external* forces. They are caused by:

   i. **ii.**

   iii.

4. Draw the free-body diagram of the (detached) collar showing all these forces labeled with their magnitudes and directions. Include any other information e.g. angles, lengths etc which may help when formulating the equations of motion.
5. The acceleration of the collar is along the bar. Show this on a kinetic diagram or on the inertial coordinate system chosen in the free-body diagram.

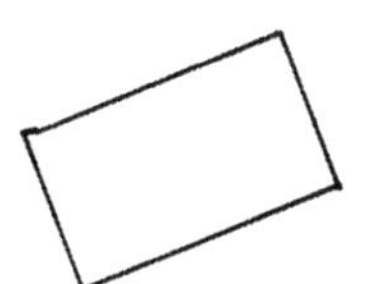

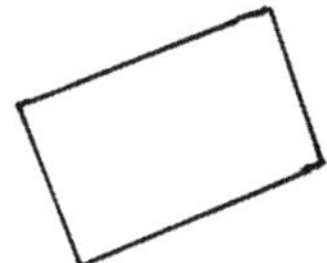

Problem 3.3

The 20 – lb collar A is initially at rest on the smooth bar. At $t = 0$ (where t is time), it is subjected to a constant force with magnitude $F = 10$ lb as shown. Draw the free-body and kinetic diagrams for the collar.

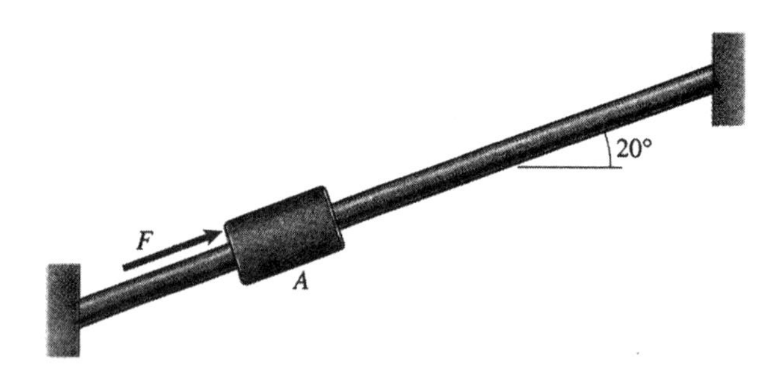

Solution

1. The size/shape of the collar does not affect the (rectilinear) motion under consideration. Consequently, we assume that the collar has *negligible size* so that it can be modelled as a particle.
2. Imagine the collar to be separated or detached from the system (collar + bar).
3. The (detached) collar is subjected to three *external* forces. They are caused by:

   i. Collar's Weight

   ii. Bar's Reaction to Collar

   iii. Force with Magnitude $F = 10$ lb
4. Draw the free-body diagram of the (detached) collar showing all these forces labeled with their magnitudes and directions. Include any other information e.g. angles, lengths etc which may help when formulating the equations of motion.
5. The acceleration of the collar is along the bar. Show this on a kinetic diagram or on the inertial coordinate system chosen in the free-body diagram.

y
x
20°
mg = 20lb
F = 10lb
N
a_x

FBD

Kinetic Diagram

Problem 3.4

Suppose in Problem 3.3, the bar is no longer smooth and the coefficients of static and kinetic friction between the collar and the bar are $\mu_s = \mu_k = 0.1$.Draw the free-body and kinetic diagrams for the collar and use them to determine if the collar moves. If the collar does move, find how fast the collar is moving and how far the collar has travelled after one second.

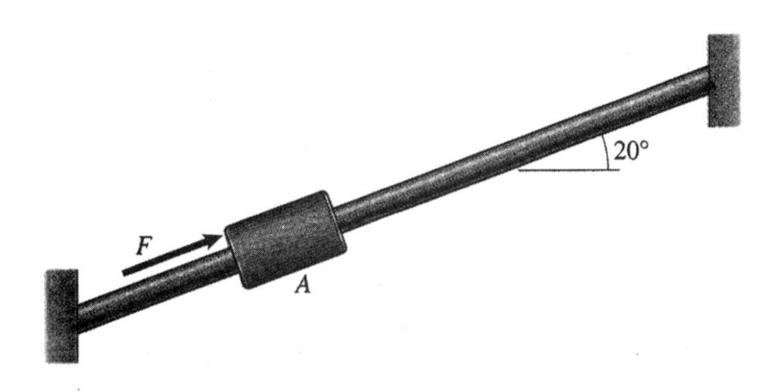

Solution

1. The size/shape of the collar does not affect any subsequent (rectilinear) motion under consideration. Consequently, we assume that the collar has *negligible size* so that it can be modelled as a particle.
2. Imagine the collar to be separated or detached from the system (collar + bar).
3. The (detached) collar is subjected to four *external* forces. They are caused by:

   i. **ii.**

   iii. **iv.**
4. Draw the free-body diagram of the (detached) collar showing all these forces labeled with their magnitudes and directions. Include any other information e.g. angles, lengths etc which may help when formulating the equations of motion.

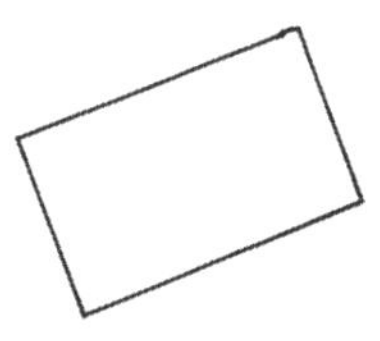

5. To determine if the collar moves, assume a static collar and determine whether the friction force necessary to keep the collar stationary is greater than the static friction force available. If this is the case, then the collar moves. Otherwise the collar remains stationary.
   Establish an appropriate $xy-$axes system such that x is parallel to the bar and y is normal to the bar. Write down the *equations of equilibrium* for the collar in the $x-$ and $y-$ directions:

   $+\rightarrow \sum F_x = ma_x :$

   $+\uparrow \sum F_y = ma_y :$

6. By solving these equations, deduce that there is insufficient (static) frictional force available to prevent the collar from moving under the given conditions

7. Using the $xy-$axes system on the free-body diagram, write down the *equations of motion* for the collar in the $x-$ and $y-$ directions:

$$+\rightarrow \sum F_x = ma_x :$$

$$+\uparrow \ \sum F_y = ma_y :$$

8. Solve for the acceleration components of the collar.

9. Integrate the acceleration components of the collar to obtain the desired speed and distance:

Problem 3.4

Suppose in Problem 3.3, the bar is no longer smooth and the coefficients of static and kinetic friction between the collar and the bar are $\mu_s = \mu_k = 0.1$.Draw the free-body and kinetic diagrams for the collar and use them to determine if the collar moves. If the collar does move, find how fast the collar is moving and how far the collar has travelled after one second.

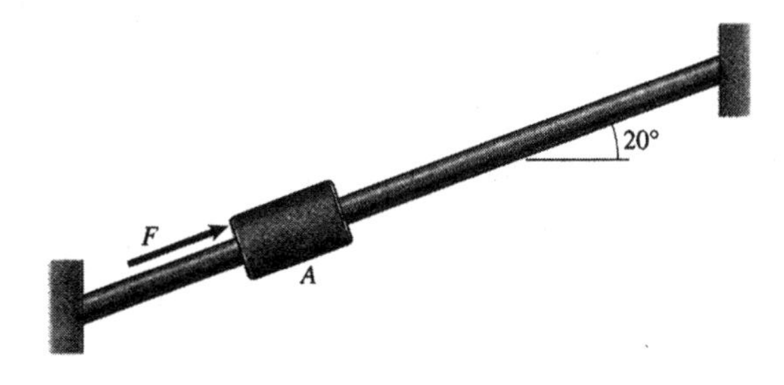

Solution

1. The size/shape of the collar does not affect any subsequent (rectilinear) motion under consideration. Consequently, we assume that the collar has *negligible size* so that it can be modelled as a particle.
2. Imagine the collar to be separated or detached from the system (collar + bar).
3. The (detached) collar is subjected to four *external* forces. They are caused by:

   i. Collar's Weight **ii. Bar's Reaction to Collar**

   iii. Force with Magnitude $F = 10$ lb **iv. Friction Force on Collar**
4. Draw the free-body diagram of the (detached) collar showing all these forces labeled with their magnitudes and directions. Include any other information e.g. angles, lengths etc which may help when formulating the equations of motion.

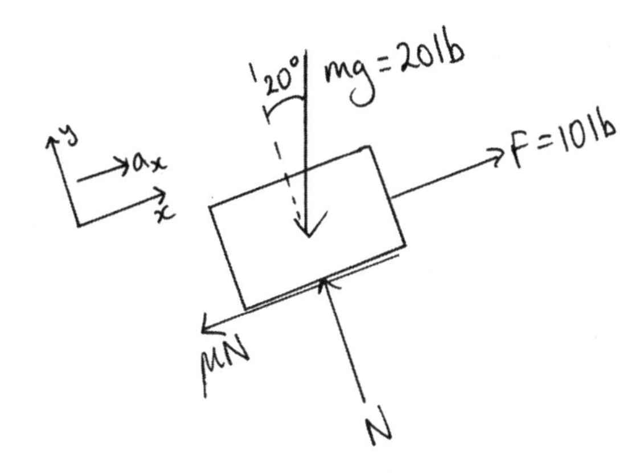

5. To determine if the collar moves, assume a static collar and determine whether the friction force necessary to keep the collar stationary is greater than the static friction force available. If this is the case, then the collar moves. Otherwise the collar remains stationary. Establish an appropriate $xy-$axes system such that x is parallel to the bar and y is normal to the bar. Write down the *equations of equilibrium* for the collar in the $x-$ and $y-$ directions:

   $+\rightarrow \sum F_x = ma_x : F - mg\sin\theta - f_{REQ} = 0 \Leftrightarrow 10 - 20\sin 20^\circ - f_{REQ} = 0$

   $+\uparrow \sum F_y = ma_y : N - mg\cos\theta = 0 \Leftrightarrow N - 20\cos 20^\circ = 0$
6. By solving these equations, deduce that there is insufficient (static) frictional force available to prevent the collar from moving under the given conditions. Solving, we obtain:

   $f_{REQ} = 3.16$ lb, $N = 18.79$ lb, and $f_{AVAIL} = \mu_S N = (0.1)(18.79) = 1.88 \text{ lb} < f_{REQ}$. Hence the collar moves.
7. Using the $xy-$axes system on the free-body diagram, write down the *equations of motion* for the collar in the $x-$ and $y-$ directions:

   $+\rightarrow \sum F_x = ma_x : F - mg\sin\theta - \mu_k N = ma_x \Leftrightarrow 10 - 20\sin 20^\circ - 0.1N = \frac{20}{g}a_x$

   $+\uparrow \sum F_y = ma_y : N - mg\cos\theta = ma_y \Leftrightarrow N - 20\cos 20^\circ = 0$

8. Solve for the acceleration components of the collar. $a_x = 2.06\ \text{ft/s}^2$
9. Integrate the acceleration components of the collar to obtain the desired speed and distance: $v_x = a_x t, x = a_x \frac{t^2}{2}$. Now let $t = 1s$ to obtain $v_x = 2.06$ ft/s, $x = 1.03$ ft.

Problem 3.5

The jeep tows the box which has a weight of 200 lb. The coefficient of kinetic friction between the box and the ground is μ_k. Draw free-body and kinetic diagrams for the box as it slides.

Solution

1. The size/shape of the box does not affect the (rectilinear) motion under consideration. Consequently, we assume that the box has *negligible size* so that it can be modelled as a particle.
2. Imagine the box to be separated or detached from the system (box + jeep + ground).
3. The (detached) box is subjected to four *external* forces. They are caused by:

   i. **ii.**

   iii. **iv.**

4. Draw the free-body diagram of the (detached) box, showing all these forces labeled with their magnitudes and directions. Include any other information e.g. angles, lengths etc which may help when formulating the equations of motion.
5. Draw the corresponding kinetic diagram.

Problem 3.5

The jeep tows the box which has a weight of 200 lb. The coefficient of kinetic friction between the box and the ground is μ_k. Draw free-body and kinetic diagrams for the box as it slides.

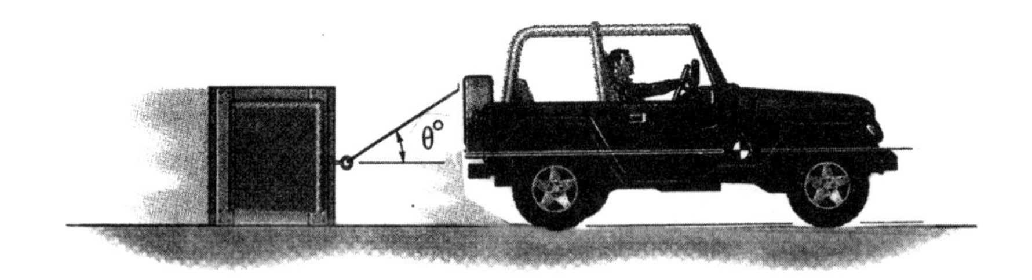

Solution

1. The size/shape of the box does not affect the (rectilinear) motion under consideration. Consequently, we assume that the box has *negligible size* so that it can be modelled as a particle.
2. Imagine the box to be separated or detached from the system (box + jeep + ground).
3. The (detached) box is subjected to four *external* forces. They are caused by:

   i. Weight of Box **ii. Towing Force in Cable**

   iii. Frictional Force **iv. Reaction from Surface**
4. Draw the free-body diagram of the (detached) box, showing all these forces labeled with their magnitudes and directions. Include any other information e.g. angles, lengths etc which may help when formulating the equations of motion.
5. Draw the corresponding kinetic diagram.

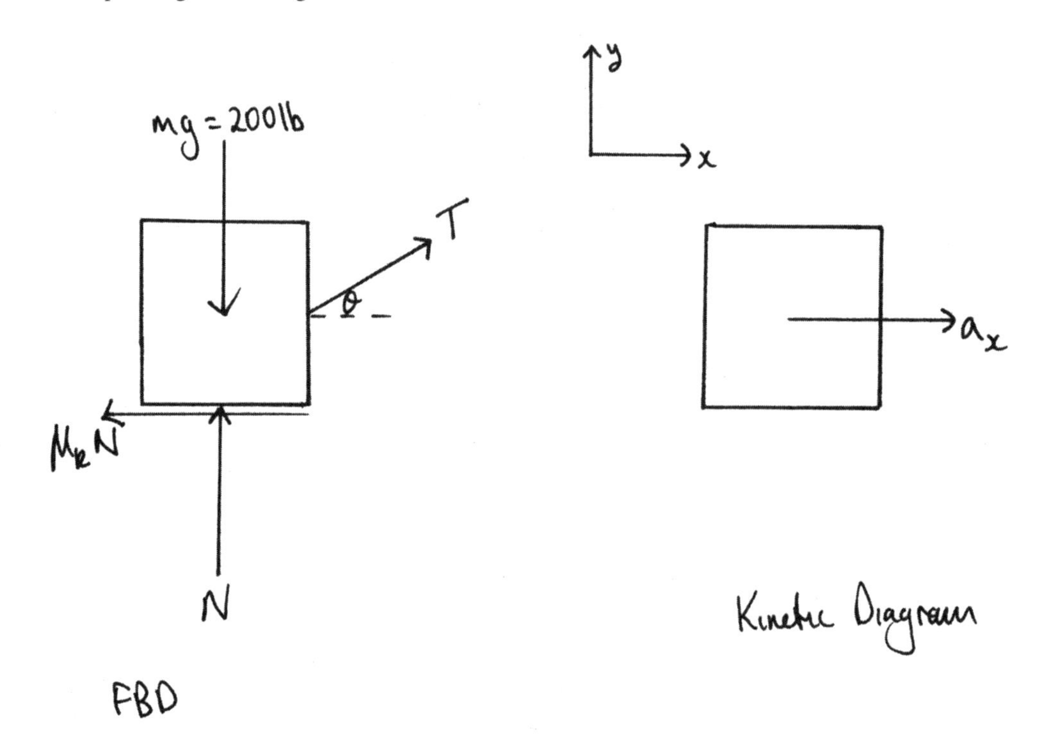

Problem 3.6

The 150 – lb person rides an elevator which has an upward acceleration a relative to the earth. Draw free-body and kinetic diagrams for the person.

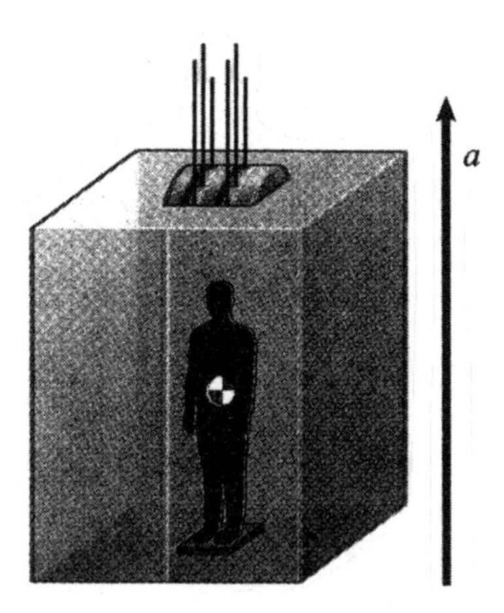

Solution

1. The size/shape of the person does not affect the (rectilinear) motion under consideration. Consequently, we assume that the person has *negligible size* and can be modelled as a particle.
2. Imagine the person to be separated or detached from the system.
3. The (detached) person is subjected to two *external* forces. They are caused by:

   i. **ii.**
4. Draw the free-body diagram of the (detached) person showing these forces labeled with their magnitudes and directions. Include any other information e.g. angles, lengths etc which may help when formulating the equations of motion.
5. Draw the corresponding kinetic diagram.

Problem 3.6

The 150−lb person rides an elevator which has an upward acceleration a relative to the earth. Draw free-body and kinetic diagrams for the person.

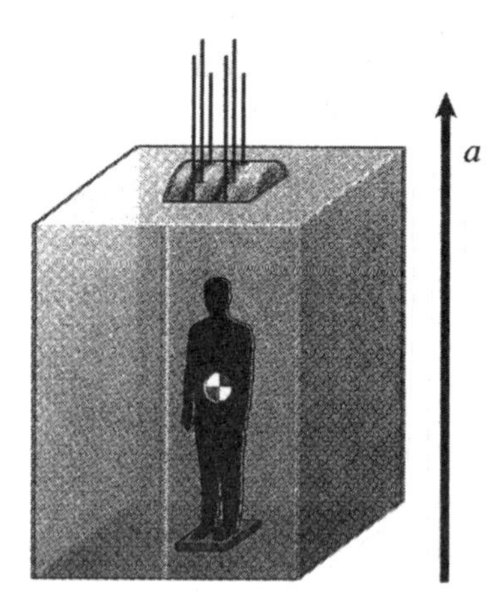

Solution

1. The size/shape of the person does not affect the (rectilinear) motion under consideration. Consequently, we assume that the person has *negligible size* and can be modelled as a particle.
2. Imagine the person to be separated or detached from the system.
3. The (detached) person is subjected to two *external* forces. They are caused by:

   i. **Person's Weight** ii. **Reaction of Supporting Surface**

4. Draw the free-body diagram of the (detached) person showing these forces labeled with their magnitudes and directions. Include any other information e.g. angles, lengths etc which may help when formulating the equations of motion.
5. Draw the corresponding kinetic diagram.

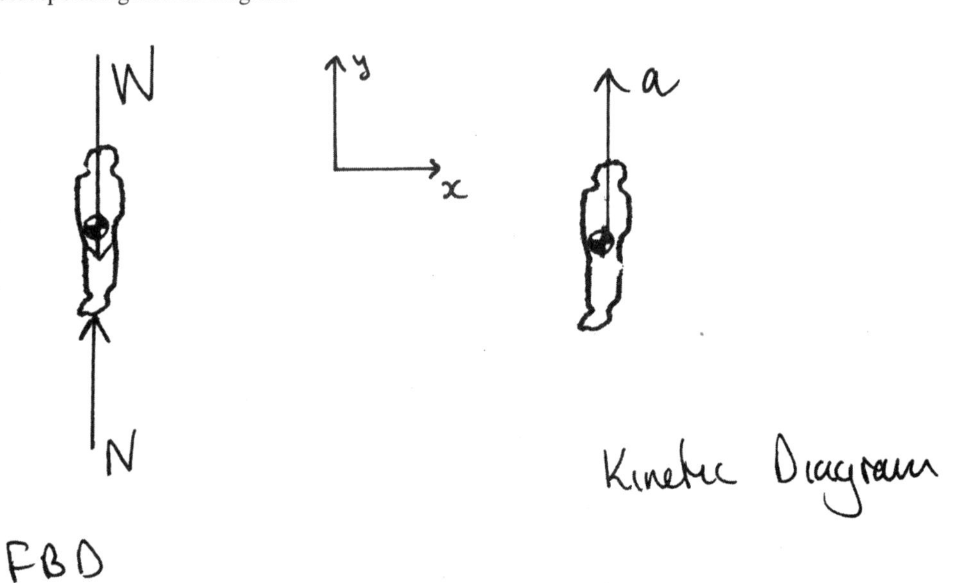

Problem 3.7

The rocket travels straight up at low altitude. Its weight at the present time is 200 kip, and the thrust of its engine is 270 kip. An onboard accelerometer indicates that the rocket's acceleration is 10 ft/s^2 upward. Draw a free-body diagram for the rocket and use it to determine the magnitude of the aerodynamic drag on the rocket.

Solution

1. The size/shape of the rocket does not affect the (rectilinear) motion under consideration. Consequently, we assume that the rocket has *negligible size* so that it can be modelled as a particle.
2. Imagine the rocket to be separated or detached from the system.
3. The (detached) rocket is subjected to three *external* forces. They are caused by:

   i. **ii.**

   iii.
4. Draw the free-body diagram of the (detached) rocket showing all these forces labeled with their magnitudes and directions. Include any other information e.g. angles, lengths etc which may help when formulating the equations of motion.
5. Draw the corresponding kinetic diagram.

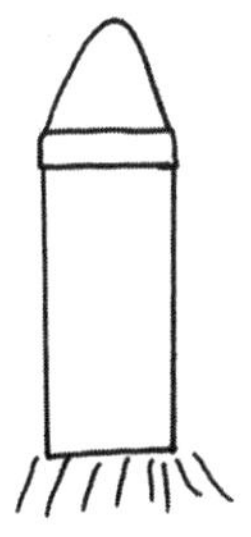

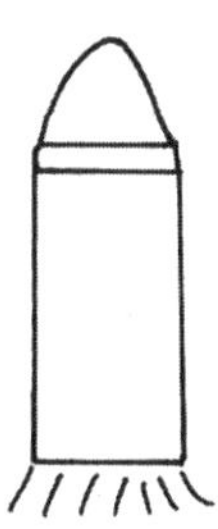

6. Using the $xy-$axes system on the free-body diagram, write down the equation of motion in the $y-$ direction:

   $+\uparrow \; \sum F_y = ma_y$:
7. Solve for the magnitude of the drag.

Problem 3.7

The rocket travels straight up at low altitude. Its weight at the present time is 200 kip, and the thrust of its engine is 270 kip. An onboard accelerometer indicates that the rocket's acceleration is 10 ft/s^2 upward. Draw a free-body diagram for the rocket and use it to determine the magnitude of the aerodynamic drag on the rocket.

Solution

1. The size/shape of the rocket does not affect the (rectilinear) motion under consideration. Consequently, we assume that the rocket has *negligible size* so that it can be modelled as a particle.
2. Imagine the rocket to be separated or detached from the system.
3. The (detached) rocket is subjected to three *external* forces. They are caused by:

   i. Rocket's Weight **ii. Thrust Force**

   iii. Drag Force
4. Draw the free-body diagram of the (detached) rocket showing all these forces labeled with their magnitudes and directions. Include any other information e.g. angles, lengths etc which may help when formulating the equations of motion.
5. Draw the corresponding kinetic diagram.

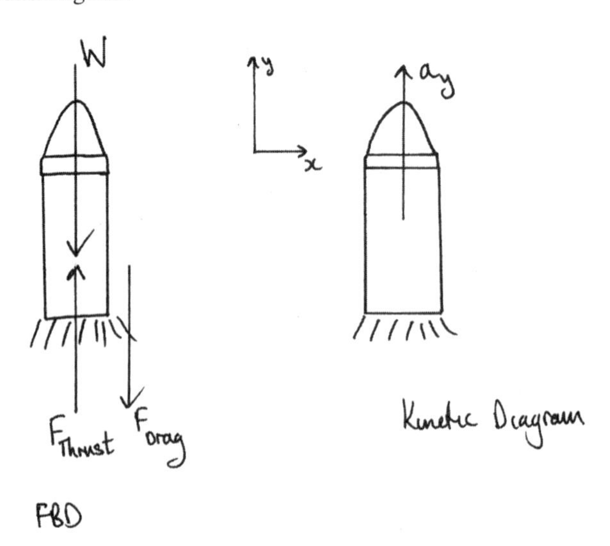

6. Using the $xy-$axes system on the free-body diagram, write down the equation of motion in the $y-$ direction:
   $+\uparrow \sum F_y = ma_y : F_{Thrust} - F_{Drag} - W = ma_y \Leftrightarrow 270000 - F_{Drag} - 200000 = \frac{20000}{32.17}\,(10)$
7. Solve for the magnitude of the drag. $F_{Drag} = 7830$ lb.

Problem 3.8

The combined weight of the motorcycle and rider is 360 lb. The coefficient of kinetic friction between the motorcycle's tires and the road is $\mu_k = 0.8$. If the rider spins the rear (drive) wheel, the normal force between the rear wheel and the road is 250 lb, and the horizontal force exerted on the front wheel by the road is negligible. Draw a free-body diagram of the combined motorcycle and rider use it to determine the resulting horizontal acceleration.

Solution

1. Imagine the combined motorcycle + rider to be separated or detached from the system.
2. The combined motorcycle + rider is subjected to three *external* forces. They are caused by

   i. **ii.**

   iii.

3. Draw the free-body diagram of the (detached) combined motorcycle and rider showing all these forces labeled with their magnitudes and directions. Include any other information e.g. angles, lengths etc which may help when formulating the equations of motion. What is the direction of the acceleration vector for the combined motorcycle and rider . Show this on a kinetic diagram or on the inertial coordinate system chosen in the free-body diagram.

4. Using the $xy-$axes system on the free-body diagram, write down the equation of motion in the (horizontal) $x-$ direction:

   $+\rightarrow \sum F_x = ma_x$:

5. Solve for the magnitude of the acceleration:

Problem 3.8

The combined weight of the motorcycle and rider is 360 lb. The coefficient of kinetic friction between the motorcycle's tires and the road is $\mu_k = 0.8$. If the rider spins the rear (drive) wheel, the normal force between the rear wheel and the road is 250 lb, and the horizontal force exerted on the front wheel by the road is negligible. Draw a free-body diagram of the combined motorcycle and rider use it to determine the resulting horizontal acceleration.

Solution

1. Imagine the combined motorcycle + rider to be separated or detached from the system.
2. The combined motorcycle + rider is subjected to three *external* forces. They are caused by

   i. Combined Weight of Motorcycle + Rider **ii. Friction**

   iii. Normal Force Between Rear Wheel and Road
3. Draw the free-body diagram of the (detached) combined motorcycle and rider showing all these forces labeled with their magnitudes and directions. Include any other information e.g. angles, lengths etc which may help when formulating the equations of motion. What is the direction of the acceleration vector for the combined motorcycle and rider . Show this on a kinetic diagram or on the inertial coordinate system chosen in the free-body diagram.

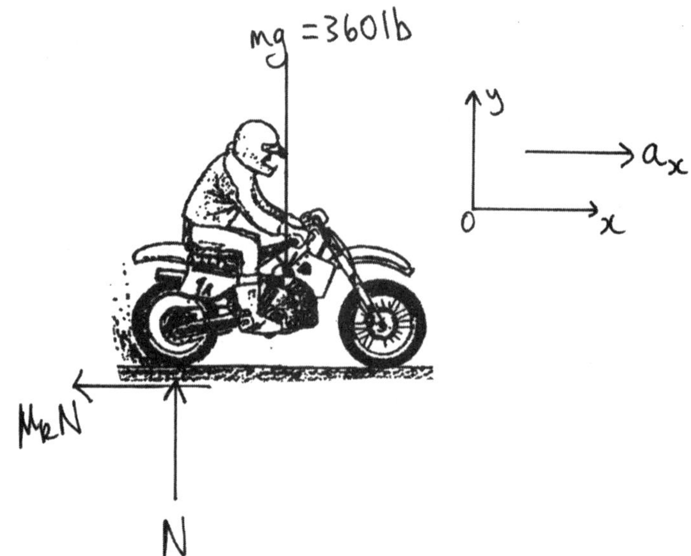

4. Using the $xy-$axes system on the free-body diagram, write down the equation of motion in the (horizontal) $x-$ direction:
   $+\rightarrow \sum F_x = ma_x : \mu_k N = ma_x \Leftrightarrow (0.8)\,250 = \frac{360}{32.17} a_x$
5. Solve for the magnitude of the acceleration: $a_x = 17.87 \text{ ft/s}^2$.

Problem 3.9

The bucket B weighs 400 lb and the acceleration of its center of mass is $\mathbf{a} = -30\mathbf{i} - 10\mathbf{j}$ (ft/sec^2). Draw free-body and kinetic diagrams for the bucket and use them to find the x and $y-$ components of the total force exerted on the bucket by its supports.

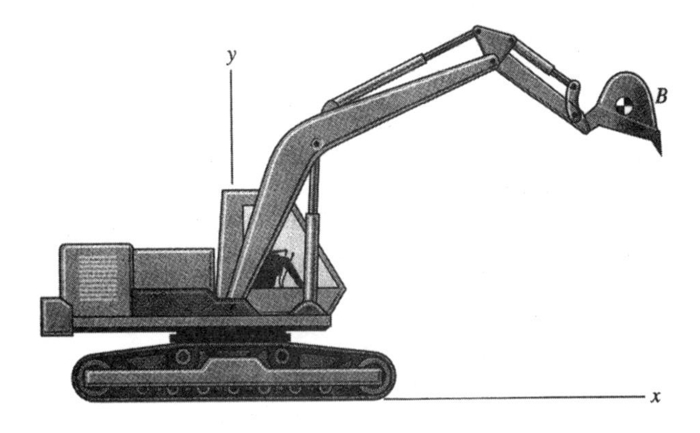

Solution

1. The bucket has *negligible size* so that it can be modelled as a particle.
2. Imagine the bucket to be separated or detached from the system.
3. The (detached) bucket is subjected to three *external* forces. They are caused by:

   i. **ii.**

   iii.
4. Draw the free-body diagram of the (detached) bucket showing all these forces labeled with their magnitudes and directions. Include any other information e.g. angles, lengths etc which may help when formulating the equations of motion. Show the corresponding acceleration components on a kinetic diagram or on the inertial coordinate system chosen in the free-body diagram.

5. Using the $xy-$axes system on the free-body diagram, write down the equation of motion in the $x-$ and $y-$ directions:

   $+\rightarrow \sum F_x = ma_x$:

   $+\uparrow \sum F_y = ma_y$:
6. Solve for the required force components.

Problem 3.9

The bucket B weighs 400 lb and the acceleration of its center of mass is $\mathbf{a} = -30\mathbf{i} - 10\mathbf{j}$ (ft/sec^2). Draw free-body and kinetic diagrams for the bucket and use them to find the x and $y-$ components of the total force exerted on the bucket by its supports.

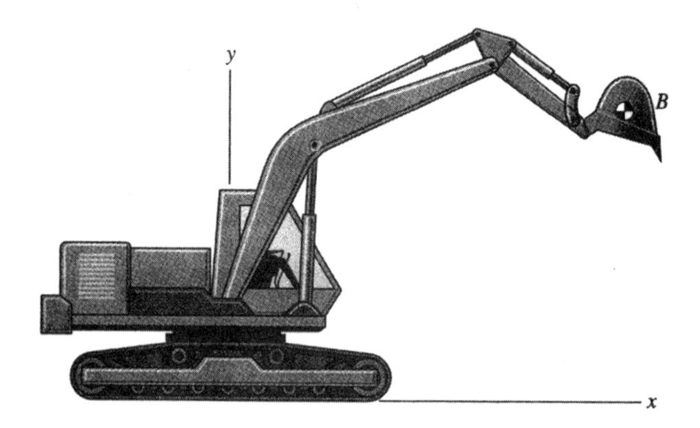

Solution

1. The bucket has *negligible size* so that it can be modelled as a particle.
2. Imagine the bucket to be separated or detached from the system.
3. The (detached) bucket is subjected to three *external* forces. They are caused by:

   i. Bucket's Weight **ii. (2) Forces Exerted on the Bucket by its Supports**

   iii.

4. Draw the free-body diagram of the (detached) bucket showing all these forces labeled with their magnitudes and directions. Include any other information e.g. angles, lengths etc which may help when formulating the equations of motion. Show the corresponding acceleration components on a kinetic diagram or on the inertial coordinate system chosen in the free-body diagram.

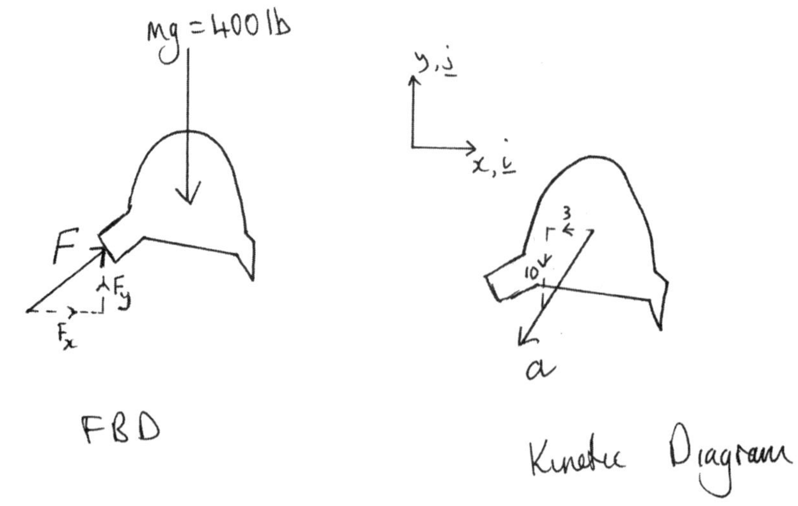

5. Using the $xy-$axes system on the free-body diagram, write down the equation of motion in the $x-$ and $y-$ directions:

   $\xrightarrow{+} \sum F_x = ma_x : F_x = ma_x \Leftrightarrow F_x = \frac{400}{32.17}(-30) = -373.0$

   $+\uparrow \sum F_y = ma_y : F_y - W = ma_y \Leftrightarrow F_y - 400 = \frac{400}{32.17}(-10)$

6. Solve for the required force components. $F_x = -373.0$ lb, $F_y = 275.7$ lb

Problem 3.10

At the instant shown, the 11000 kg airplane is subjected to the thrust $T = 110$ kN, lift $L = 260$ kN and the drag $D = 34$ kN. Use a free-body diagram of the airplane to determine the magnitude of its acceleration.

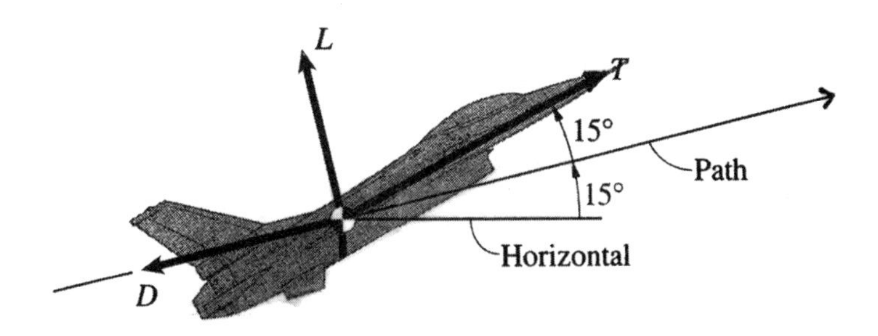

Solution

1. Imagine the airplane to be separated or detached from the system.
2. The (detached) airplane is subjected to four *external* forces. They are caused by:

   i. **ii.**

   iii. **iv.**

3. Draw the free-body diagram of the (detached) airplane showing all these forces labeled with their magnitudes and directions. Include any other information e.g. angles, lengths etc which may help when formulating the equations of motion. Show the corresponding acceleration components on a kinetic diagram or on the inertial coordinate system chosen in the free-body diagram

4. Using the xy–axes system on the free-body diagram, write down the equations of motion in the x and y– directions:

   $\sum F_x = ma_x$:

   $\sum F_y = ma_y$:
5. Solve for the required acceleration components and hence the magnitude of the acceleration of the aircraft.

Problem 3.10

At the instant shown, the 11000 kg airplane is subjected to the thrust $T = 110$ kN, lift $L = 260$ kN and the drag $D = 34$ kN. Use a free-body diagram of the airplane to determine the magnitude of its acceleration.

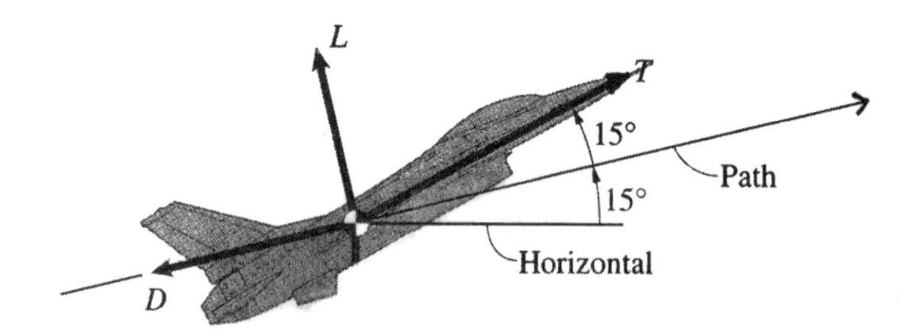

Solution

1. Imagine the airplane to be separated or detached from the system.
2. The (detached) airplane is subjected to four *external* forces. They are caused by:
   - **i. Thrust Force**
   - **ii. Lift Force**
   - **iii. Airplane's Weight**
   - **iv. Drag Force**
3. Draw the free-body diagram of the (detached) airplane showing all these forces labeled with their magnitudes and directions. Include any other information e.g. angles, lengths etc which may help when formulating the equations of motion. Show the corresponding acceleration components on a kinetic diagram or on the inertial coordinate system chosen in the free-body diagram

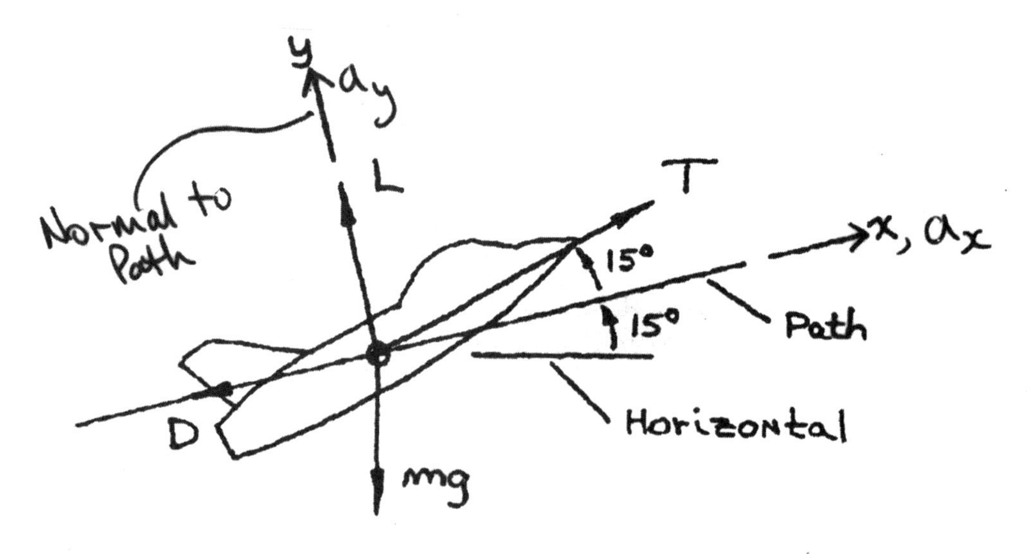

4. Using the $xy-$axes system on the free-body diagram, write down the equations of motion in the x and $y-$ directions:

   $\sum F_x = ma_x : T \cos 15^\circ - D - mg \sin 15^\circ = ma_x$

   $\sum F_y = ma_y : L + T \sin 15^\circ - mg \cos 15^\circ = ma_y$
5. Solve for the required acceleration components and hence the magnitude of the acceleration of the aircraft.

   $a_x = 4.03 \ \text{m/s}^2, a_y = 16.75 \ \text{m/s}^2, |\mathbf{a}| = \sqrt{a_x^2 + a_y^2} = 17.23 \ \text{m/s}^2.$

Problem 3.11

Each box weighs 50 lb. Draw the free-body and kinetic diagrams for each box as they slide from their initial positions (the coefficient of kinetic friction between the boxes and the surface is μ_k).

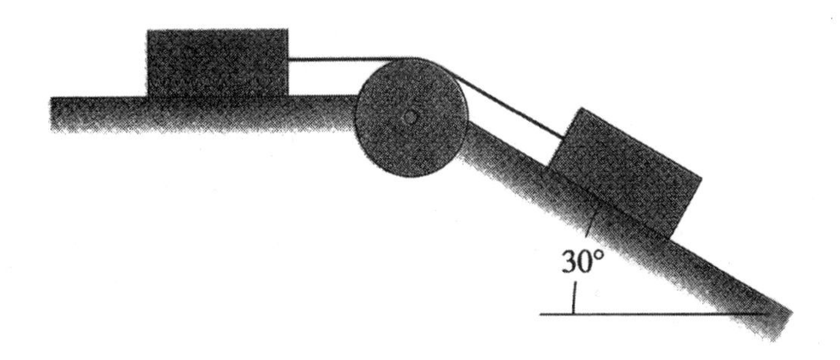

Solution

1. The size/shape of the each box does not affect the (rectilinear) motion under consideration. Consequently, we assume that the boxes have *negligible size* so that each can be modelled as a particle.
2. Imagine each box to be separated or detached from the system.
3. Each box is subjected to four *external* forces.
4. Draw the free-body diagram of each (detached) box showing all these forces labeled with their magnitudes and directions. Include any other information e.g. angles, lengths etc which may help when formulating the equations of motion.
5. Show the acceleration of each block on a kinetic diagram or on the inertial coordinate system chosen in the corresponding free-body diagram.

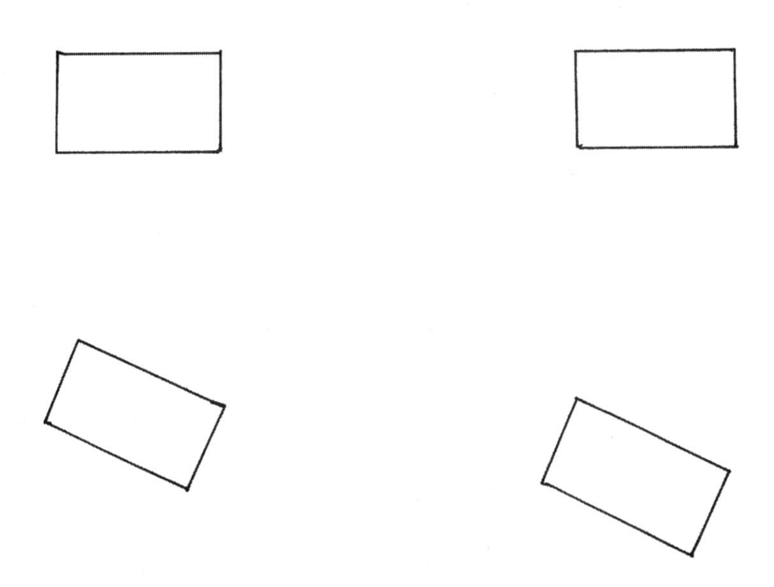

Problem 3.11

Each box weighs 50 lb. Draw the free-body and kinetic diagrams for each box as they slide from their initial positions (the coefficient of kinetic friction between the boxes and the surface is μ_k).

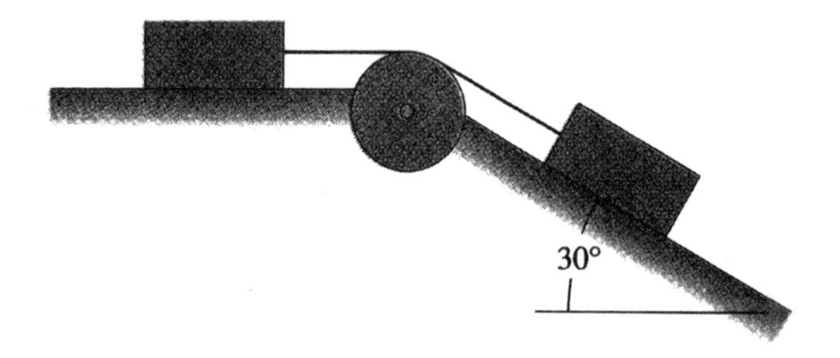

Solution

1. The size/shape of the each box does not affect the (rectilinear) motion under consideration. Consequently, we assume that the boxes have *negligible size* so that each can be modelled as a particle.
2. Imagine each box to be separated or detached from the system.
3. Each box is subjected to four *external* forces.
4. Draw the free-body diagram of each (detached) box showing all these forces labeled with their magnitudes and directions. Include any other information e.g. angles, lengths etc which may help when formulating the equations of motion.
5. Show the acceleration of each block on a kinetic diagram or on the inertial coordinate system chosen in the corresponding free-body diagram.

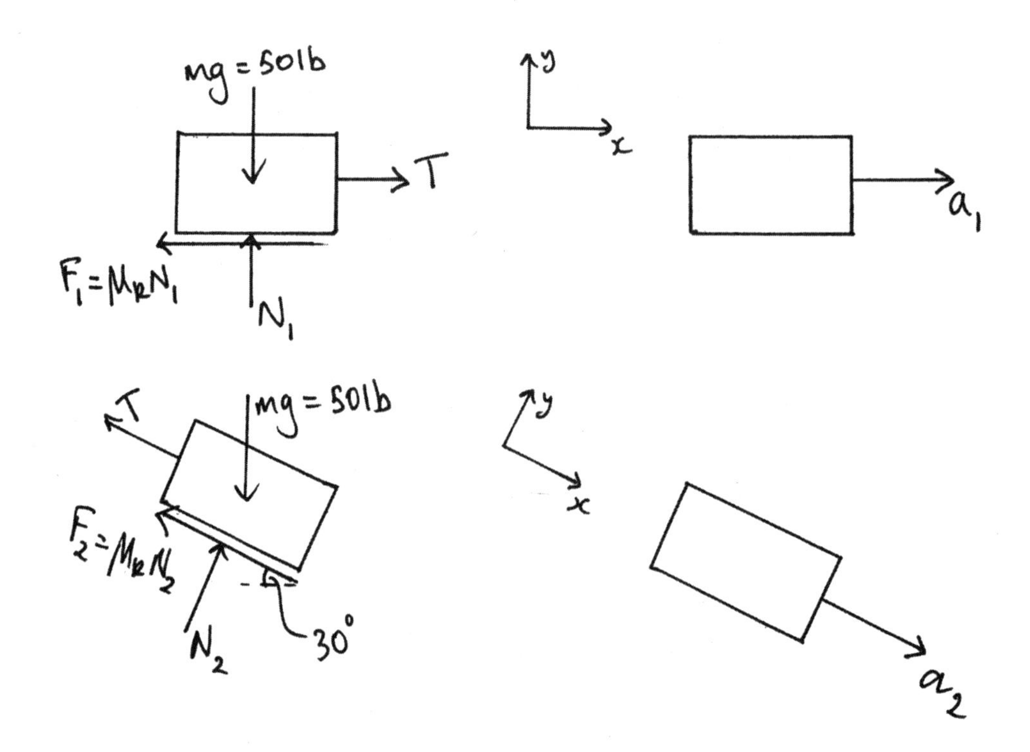

Problem 3.12

The boat weighs 2600 lb with its passengers. It is moving in a circular path of radius $R = 80$ ft at a constant speed of 15 mi/hr. Draw free-body diagrams of the boat from both top and rear viewpoints, showing forces acting in all three coordinate directions. Use these diagrams to determine the *total* horizontal force on the boat in the direction tangent to its path, the *total* horizontal force on the boat in the direction perpendicular to its path and the *total* vertical force acting on the boat.

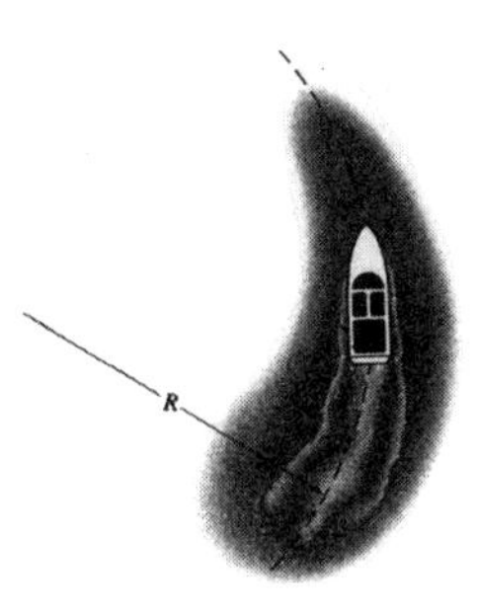

Solution

1. The boat has *negligible size* so that it can be modelled as a particle.
2. Imagine the boat to be separated or detached from the system.
3. The (detached) boat is subjected to four *external* forces.
4. Draw the free-body diagrams of the (detached) boat from both the top and rear viewpoints. Which information given in the question suggests you use a $nt-$coordinate system as the chosen inertial system?

   Show the corresponding acceleration components on a kinetic diagram or on the inertial coordinate system chosen in the free-body diagrams.

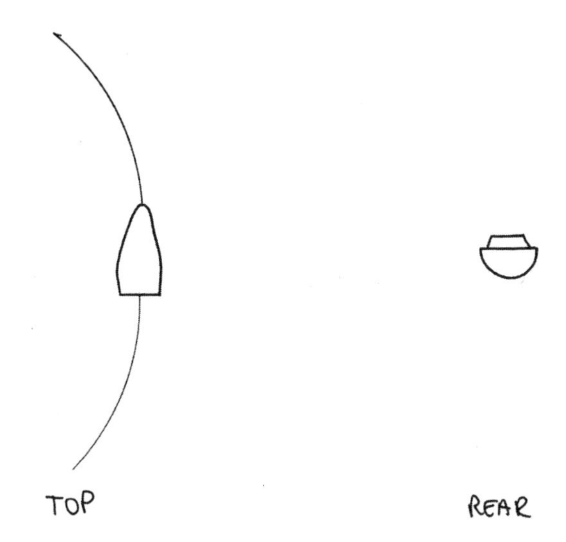

5. Using the $nt-$axes system on the free-body diagram, write down the equation of motion in the $n-$ and $t-$directions:

   $+ \leftarrow \sum F_x = ma_x$:

   $+\uparrow \ \sum F_y = ma_y$:

6. Write down the equation of motion in the vertical direction:

$$+\uparrow \ \sum F_{vert} = ma_{vert} :$$

7. Solve for the required *total* forces.

Problem 3.12

The boat weighs 2600 lb with its passengers. It is moving in a circular path of radius $R = 80$ ft at a constant speed of 15 mi/hr. Draw free-body diagrams of the boat from both top and rear viewpoints, showing forces acting in all three coordinate directions. Use these diagrams to determine the *total* horizontal force on the boat in the direction tangent to its path, the *total* horizontal force on the boat in the direction perpendicular to its path and the *total* vertical force acting on the boat.

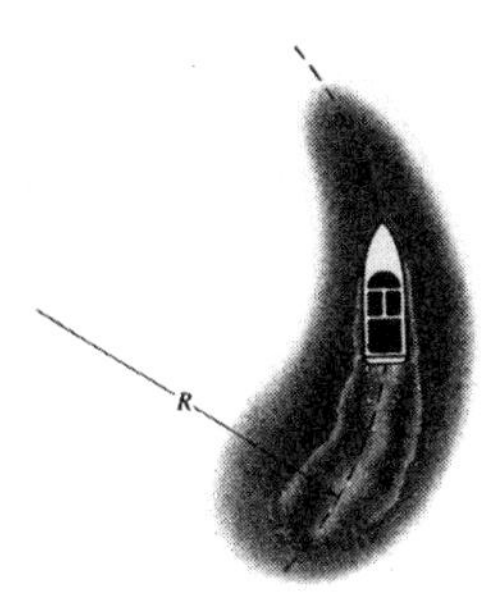

Solution

1. The boat has *negligible size* so that it can be modelled as a particle.
2. Imagine the boat to be separated or detached from the system.
3. The (detached) boat is subjected to four *external* forces.
4. Draw the free-body diagrams of the (detached) boat from both the top and rear viewpoints. Which information given in the question suggests you use a $nt-$coordinate system as the chosen inertial system?
   CURVILINEAR PATH
   Show the corresponding acceleration components on a kinetic diagram or on the inertial coordinate system chosen in the free-body diagrams.

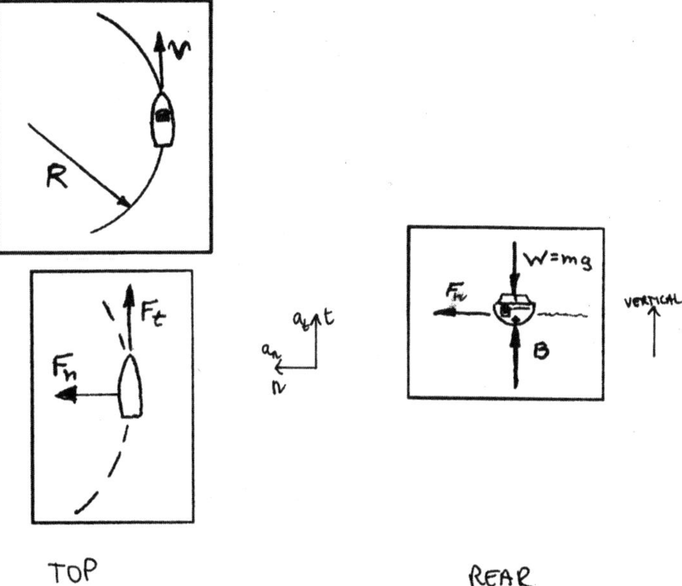

5. Using the $nt-$axes system on the free-body diagram, write down the equation of motion in the $n-$ and $t-$directions:

$$+\leftarrow \sum F_x = ma_x : \sum F_n = \frac{mv^2}{R}$$

$$+\uparrow \ \sum F_y = ma_y : \sum F_t = m\frac{dv}{dt}$$

6. Write down the equation of motion in the vertical direction: $+\uparrow \ \sum F_{vert} = ma_{vert} : B - mg = 0$

7. Solve for the required *total* forces. Here, $\frac{dv}{dt} = 0$, $R = 80$ ft, $v = 15$ mi/hr $= 22$ ft/s.
The equations of motion now give $a_t = 0$, $a_n = 6.05\ \text{ft/s}^2$, $a_{vert} = 0$. Hence, the associated forces are:

$$\sum F_t = 0,\ \sum F_n = 488.5 \text{ lb (inward, toward center of curvature), .}$$
$$\sum F_{vert} = 0.$$

Problem 3.13

Each of the blocks has mass m. The coefficient of kinetic friction at all surfaces of contact is μ_k. A horizontal force **R** is applied to the bottom block. Draw free-body diagrams for each of the top and bottom blocks.

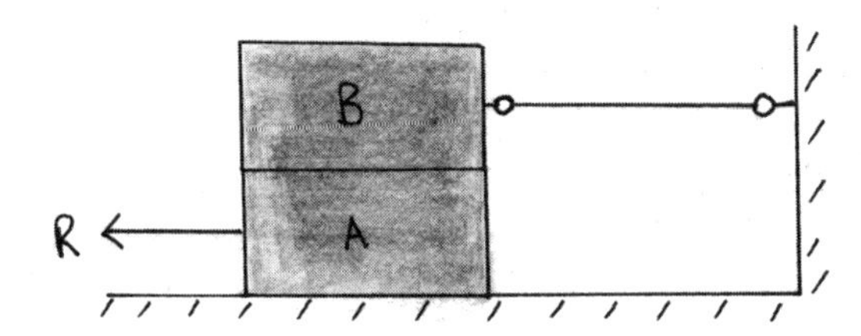

Solution

1. The size/shape of the blocks does not affect the motion under consideration. Consequently, we assume that the blocks have *negligible size* so that they can be modelled as particles.
2. Imagine each block to be separated or detached from the system (two blocks + plane).
3. The (detached) upper block is subjected to four *external* forces. They are caused by:
   The (detached) lower block is subjected to six external forces. They are caused by:

i.	**ii.**
iii.	**iv.**
v.	**vi.**

4. Draw the free-body diagrams of each (detached) block showing all these forces labeled with their magnitudes and directions. Include any other information e.g. angles, lengths etc which may help when formulating the equations of motion.
5. Show the acceleration of block A on a kinetic diagram or on the inertial coordinate system chosen in the free-body diagram.

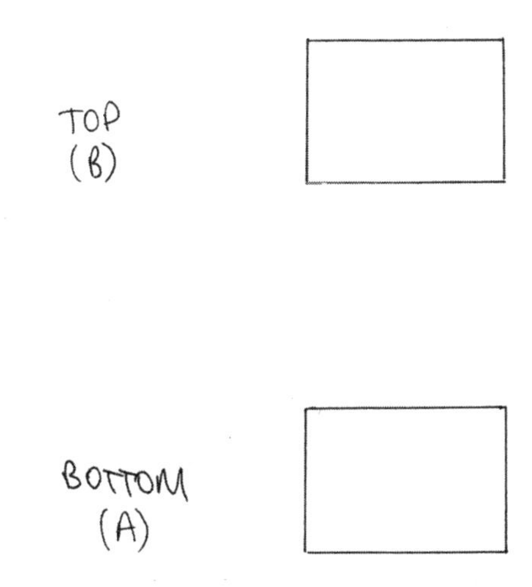

Problem 3.13

Each of the blocks has mass m. The coefficient of kinetic friction at all surfaces of contact is μ_k. A horizontal force **R** is applied to the bottom block. Draw free-body diagrams for each of the top and bottom blocks.

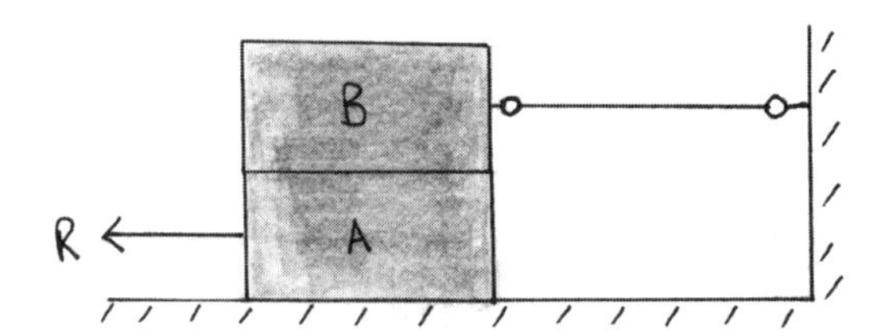

Solution

1. The size/shape of the blocks does not affect the motion under consideration. Consequently, we assume that the blocks have *negligible size* so that they can be modelled as particles.
2. Imagine each block to be separated or detached from the system (two blocks + plane).
3. The (detached) upper block is subjected to four *external* forces. They are caused by:

   i. It's Weight **ii. CableTension T**

   iii. Friction Between Blocks **iv. Reaction From Lower Block**

   The (detached) lower block is subjected to six external forces. They are caused by:

   i. It's Weight **ii. Force P**

   iii. Friction at Supporting Surface **iv. Friction with Upper Block**

   v. Reaction From Supporting Surface **vi. Reaction from Upper Block**
4. Draw the free-body diagrams of each (detached) block showing all these forces labeled with their magnitudes and directions. Include any other information e.g. angles, lengths etc which may help when formulating the equations of motion.
5. Show the acceleration of block A on a kinetic diagram or on the inertial coordinate system chosen in the free-body diagram.

TOP (B)

mg

T

y x

$\mu_k N_L$

$N_L = mg$

($N_L = mg$ from equilibrium in y-direction)

$N_L = mg$

$\mu_k N_L$

P

BOTTOM (A)

mg

a y x

$\mu_k N_s$

N_s

($N_s = 2mg$ from equilibrium in y-direction)

Problem 3.14

Block A has mass m_1 and block B has mass m_2. The coefficient of kinetic friction at all surfaces of contact is μ_k. Draw free-body and kinetic diagrams for each of the blocks. Explain the significance of each quantity on the free-body and kinetic diagrams.

Solution

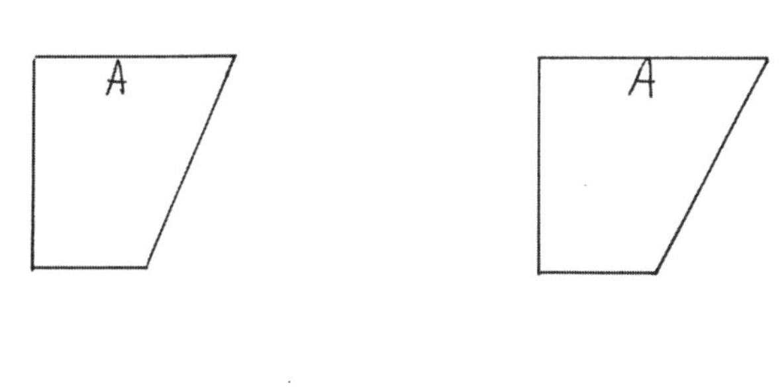

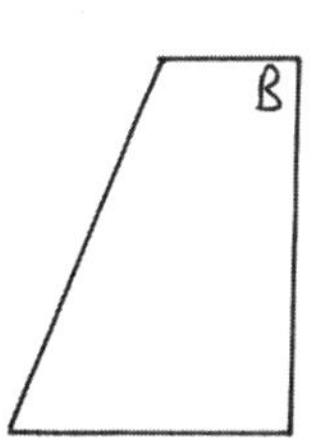

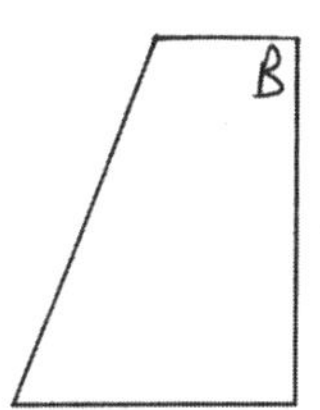

Problem 3.14

Block A has mass m_1 and block B has mass m_2. The coefficient of kinetic friction at all surfaces of contact is μ_k. Draw free-body and kinetic diagrams for each of the blocks. Explain the significance of each quantity on the free-body and kinetic diagrams.

Solution

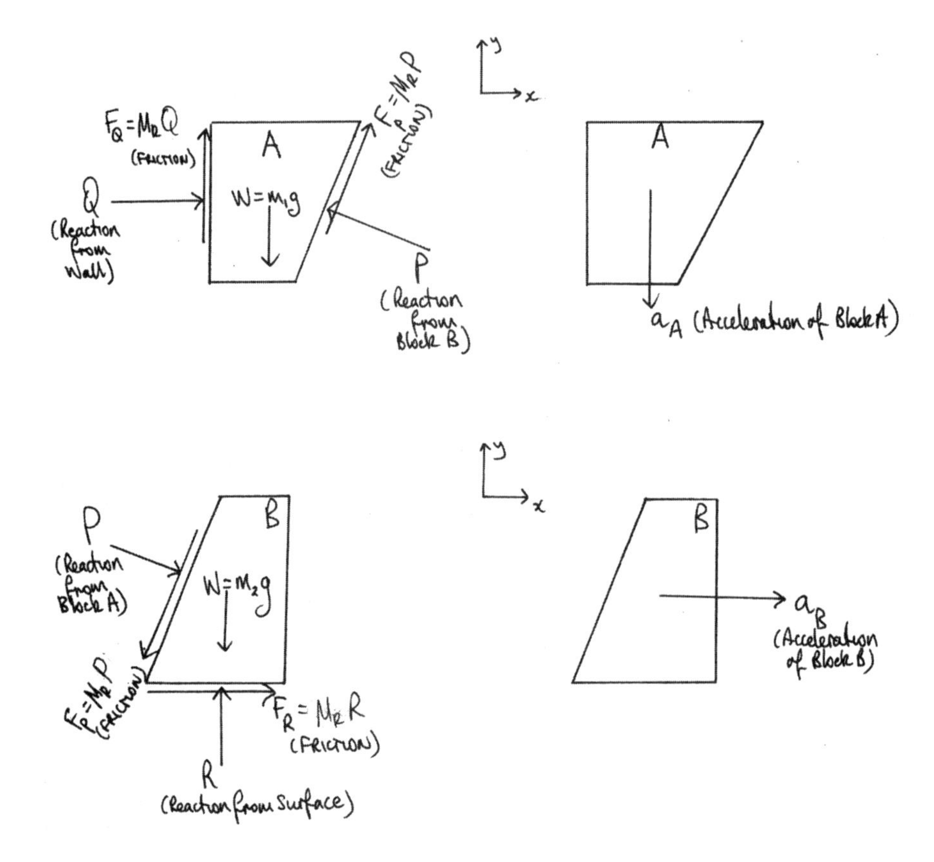

Problem 3.15

Block A has mass m_1 and block B has mass m_2. The coefficient of kinetic friction at all surfaces of contact is μ_k. Draw free-body and kinetic diagrams for each of the blocks (assume that sliding has begun). Explain the significance of each quantity on the free-body and kinetic diagrams.

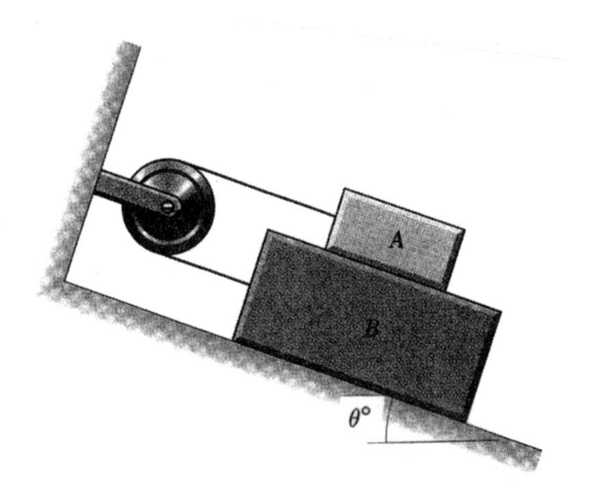

Solution

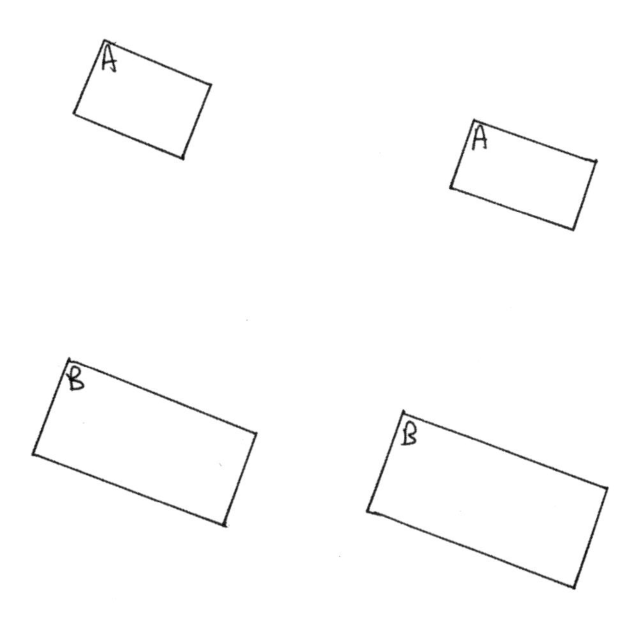

Problem 3.15

Block A has mass m_1 and block B has mass m_2. The coefficient of kinetic friction at all surfaces of contact is μ_k. Draw free-body and kinetic diagrams for each of the blocks (assume that sliding has begun). Explain the significance of each quantity on the free-body and kinetic diagrams.

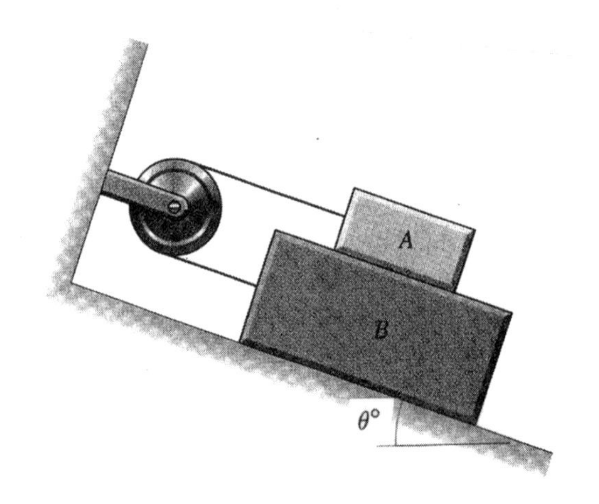

Solution

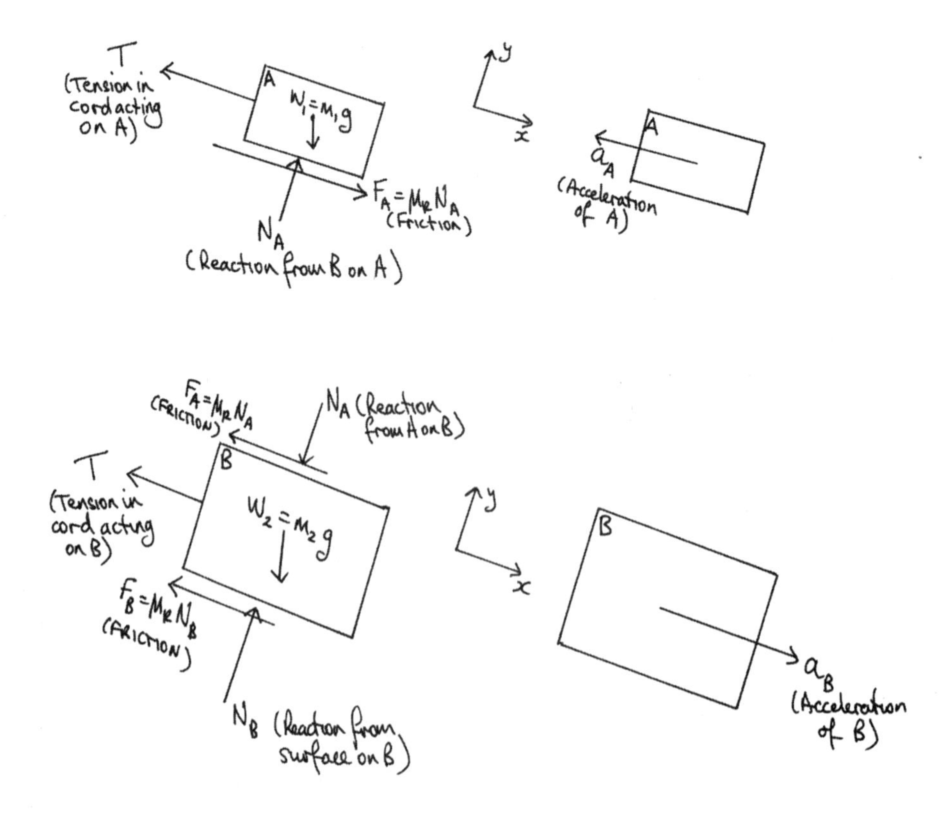

3.2 Free-Body Diagrams in Rigid Body Kinetics

Problem 3.16

The refrigerator of mass m rests on casters at A and B. Suppose that you push on it with a horizontal force of magnitude F as shown and that the casters remain on the floor. Draw the free-body and kinetic diagrams of the refrigerator. Neglect the mass of the casters.

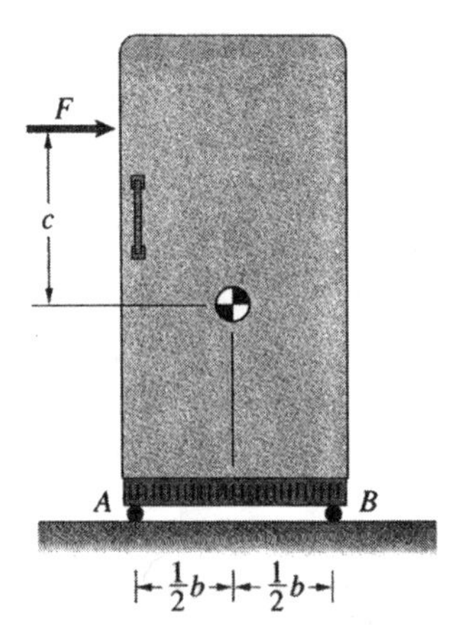

Solution

1. Imagine the refrigerator to be separated or detached from the system (refrigerator + floor).
2. The refrigerator is subjected to four *external* forces. They are caused by:

   i. **ii.**

   iii. **iv.**

3. Draw the free-body diagram of the (detached) refrigerator showing all these forces labeled with their magnitudes and directions. Include any other relevant information e.g. lengths, angles etc which may help when formulating the equations of motion (including the moment equation) for the refrigerator.
4. Draw the corresponding kinetic diagram.

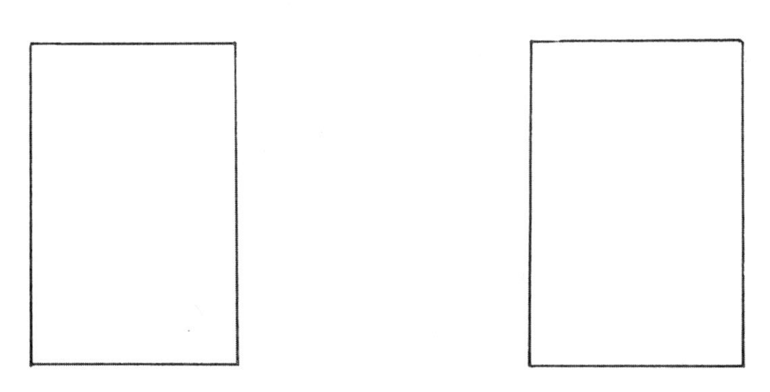

Problem 3.16

The refrigerator of mass m rests on casters at A and B. Suppose that you push on it with a horizontal force of magnitude F as shown and that the casters remain on the floor. Draw the free-body and kinetic diagrams of the refrigerator. Neglect the mass of the casters.

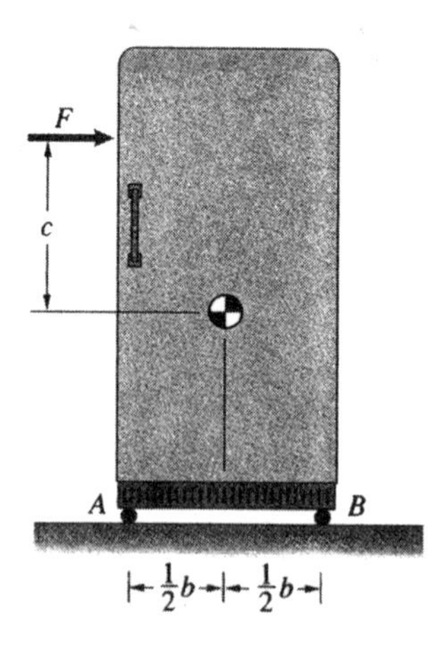

Solution

1. Imagine the refrigerator to be separated or detached from the system (refrigerator + floor).
2. The refrigerator is subjected to four *external* forces. They are caused by:
   - **i. Force F**
   - **ii. Weight**
   - **iii. Reaction at** A
   - **iv. Reaction at** B
3. Draw the free-body diagram of the (detached) refrigerator showing all these forces labeled with their magnitudes and directions. Include any other relevant information e.g. lengths, angles etc which may help when formulating the equations of motion (including the moment equation) for the refrigerator.
4. Draw the corresponding kinetic diagram.

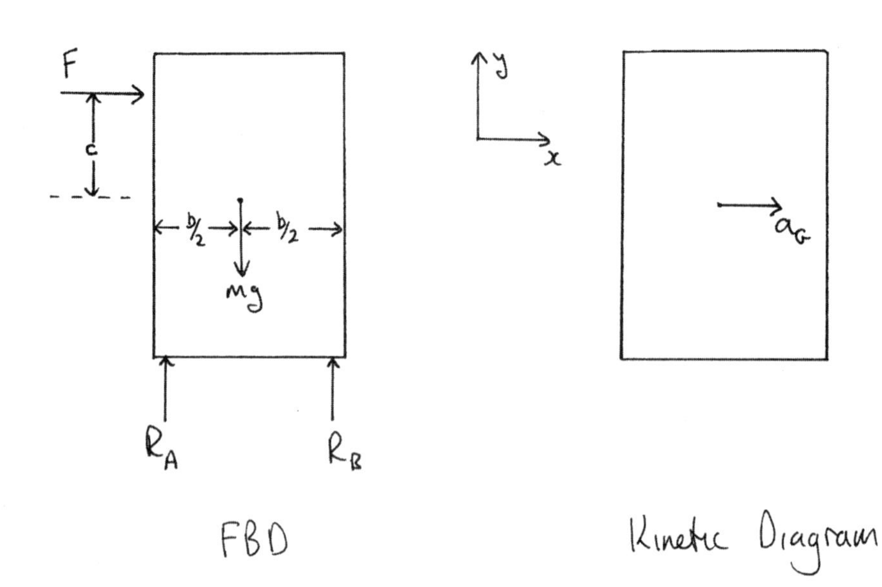

Problem 3.17

The 14000 lb airplane's arresting hook exerts the force of magnitude F and causes the plane to decelerate at $6g$. Draw the free-body and kinetic diagrams of the airplane. Neglect the mass of the wheels.

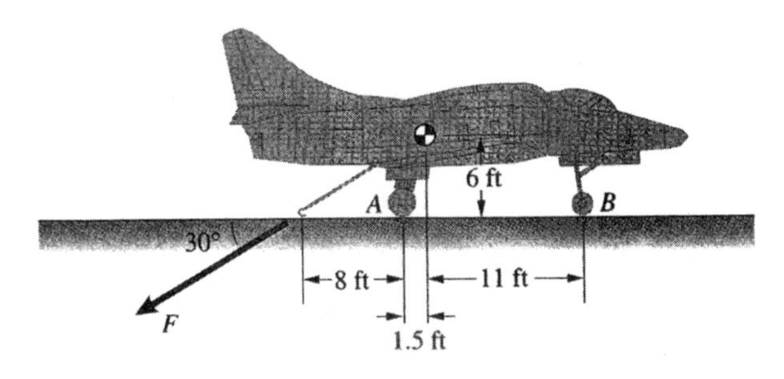

Solution

1. Imagine the airplane to be separated or detached from the system (airplane + ground).
2. The airplane is subjected to six *external* forces. They are caused by:

   i.   **ii.**

   iii.   **iv.**

   v.   **vi.**

3. Draw the free-body diagram of the (detached) airplane showing all these forces labeled with their magnitudes and directions. Include any other relevant information e.g. lengths, angles etc which may help when formulating the equations of motion (including the moment equation) for the airplane.
4. Draw the corresponding kinetic diagram.

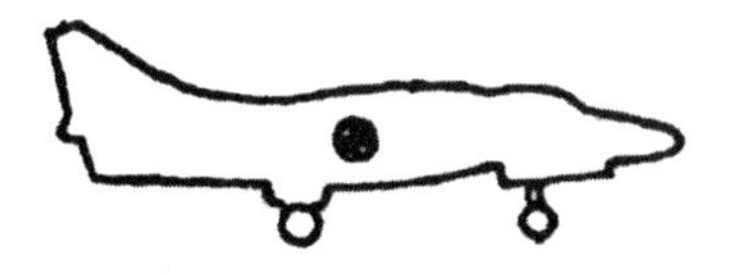

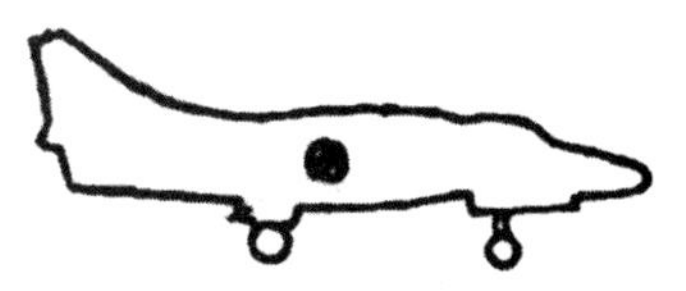

Problem 3.17

The 14000 lb airplane's arresting hook exerts the force of magnitude F and causes the plane to decelerate at $6g$. Draw the free-body and kinetic diagrams of the airplane. Neglect the mass of the wheels.

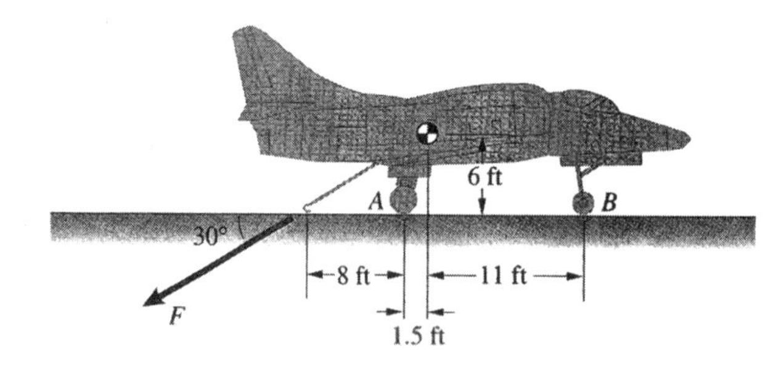

Solution

1. Imagine the airplane to be separated or detached from the system (airplane + ground).
2. The airplane is subjected to six *external* forces. They are caused by:
   - i. **Weight**
   - ii. **Force F**
   - iii. **Reaction at** *A*
   - iv. **Reaction at** *B*
   - v. **Friction at** *A*
   - vi. **Friction at** *B*
3. Draw the free-body diagram of the (detached) airplane showing all these forces labeled with their magnitudes and directions. Include any other relevant information e.g. lengths, angles etc which may help when formulating the equations of motion (including the moment equation) for the airplane.
4. Draw the corresponding kinetic diagram.

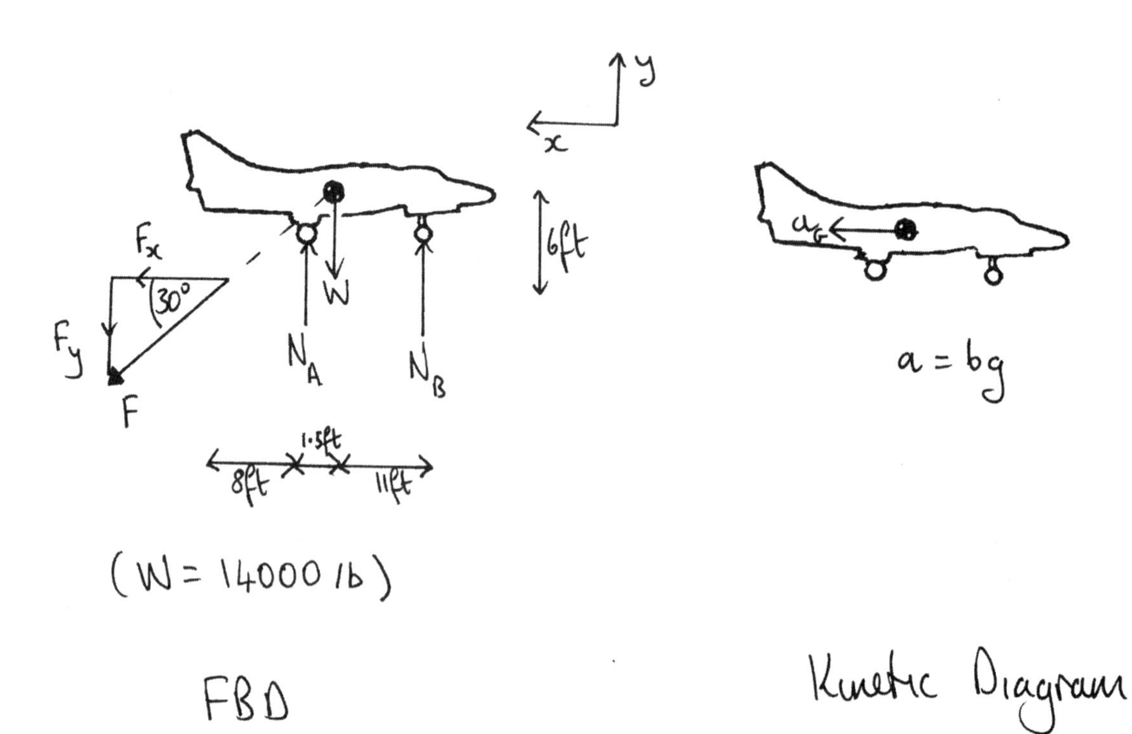

Problem 3.18

A student catching a ride to his summer job unwisely supports himself in the back of an accelerating truck by exerting a horizontal force of magnitude F_H on the truck's cab at A. Draw the free-body and kinetic diagrams of the student.

Solution

1. Imagine the student to be separated or detached from the system.
2. The student is subjected to four *external* forces. They are caused by:

   i.   **ii.**

   iii.   **iv.**

3. Draw the free-body diagram of the (detached) student showing all these forces labeled with their magnitudes and directions. Include any other relevant information e.g. lengths, angles etc which may help when formulating the equations of motion (including the moment equation) for the student.
4. Draw the corresponding kinetic diagram.

Problem 3.18

A student catching a ride to his summer job unwisely supports himself in the back of an accelerating truck by exerting a horizontal force of magnitude F_H on the truck's cab at A. Draw the free-body and kinetic diagrams of the student.

Solution

1. Imagine the student to be separated or detached from the system.
2. The student is subjected to four *external* forces. They are caused by:
   - **i. His weight**
   - **ii. Normal force on his feet**
   - **iii. Friction on his feet**
   - **iv. Force F_H exerted on student by truck**
3. Draw the free-body diagram of the (detached) student showing all these forces labeled with their magnitudes and directions. Include any other relevant information e.g. lengths, angles etc which may help when formulating the equations of motion (including the moment equation) for the student.
4. Draw the corresponding kinetic diagram.

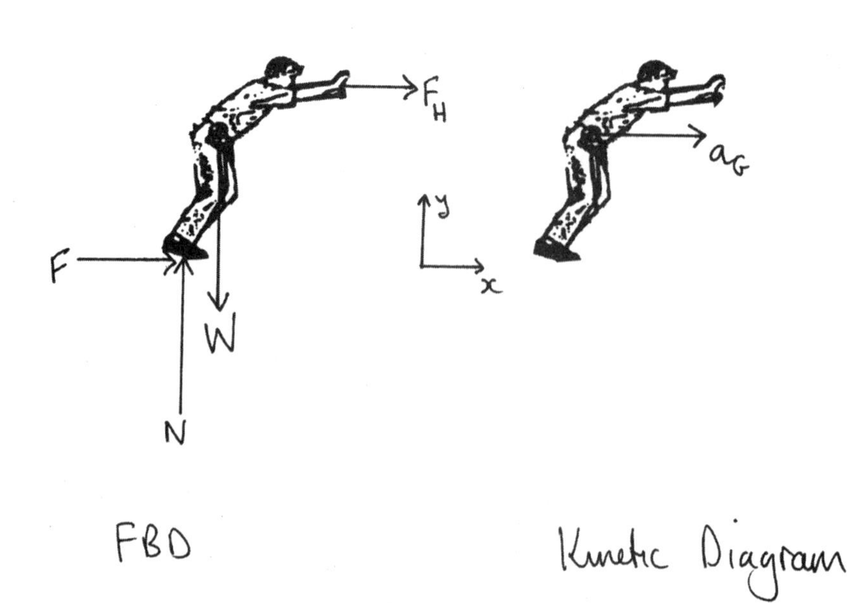

Problem 3.19

The combined mass of the person and bicycle is m. The location of their combined center of mass is as shown. Draw the free-body and kinetic diagrams of the combined person and bicycle.

Solution

1. Imagine the combined person and bicycle to be separated or detached from the system.
2. The combined person and bicycle is subjected to five *external* forces. They are caused by:

   i. **ii.**

   iii. **iv.**

   v.

3. Draw the free-body diagram of the (detached) combined person and bicycle showing all these forces labeled with their magnitudes and directions. Include any other relevant information e.g. lengths, angles etc which may help when formulating the equations of motion (including the moment equation) for the pipe.
4. Draw the corresponding kinetic diagram.

Problem 3.19

The combined mass of the person and bicycle is m. The location of their combined center of mass is as shown. Draw the free-body and kinetic diagrams of the combined person and bicycle.

Solution

1. Imagine the combined person and bicycle to be separated or detached from the system.
2. The combined person and bicycle is subjected to five *external* forces. They are caused by:
   - i. **Weight**
   - ii. **Reaction at** A
   - iii. **Reaction at** B
   - iv. **Friction at** A
   - v. **Friction at** B
3. Draw the free-body diagram of the (detached) combined person and bicycle showing all these forces labeled with their magnitudes and directions. Include any other relevant information e.g. lengths, angles etc which may help when formulating the equations of motion (including the moment equation) for the pipe.
4. Draw the corresponding kinetic diagram.

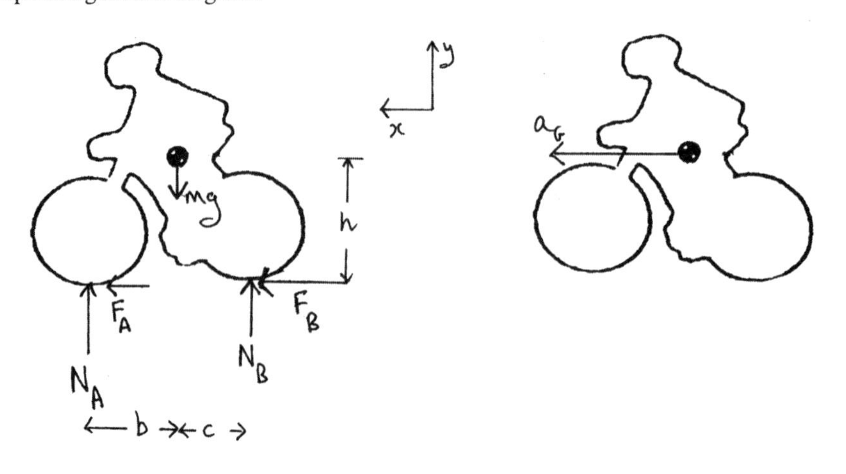

FBD

Kinetic Diagram

Problem 3.20

The slender bar has mass m and is released from rest in the horizontal position. Draw the free-body and kinetic diagrams of the bar at that instant.

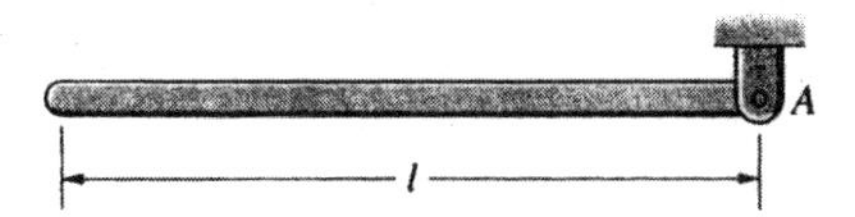

Solution

1. Imagine the bar to be separated or detached from the system.
2. The desk is subjected to three *external* forces. They are caused by:

   i.   **ii.**

   iii.

3. Draw the free-body diagram of the (detached) bar showing all these forces labeled with their magnitudes and directions. Include any other relevant information e.g. lengths, angles etc which may help when formulating the equations of motion (including the moment equation) for the bar.
4. Draw the corresponding kinetic diagram for the bar at the instant in question.

Problem 3.20

The slender bar has mass m and is released from rest in the horizontal position. Draw the free-body and kinetic diagrams of the bar at that instant.

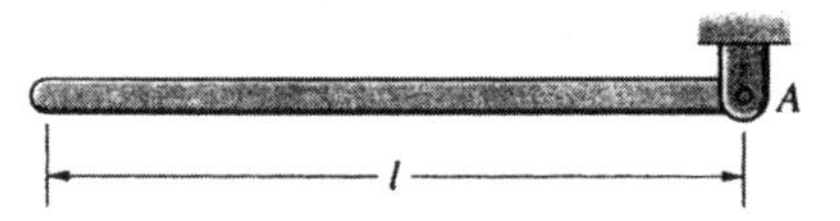

Solution

1. Imagine the bar to be separated or detached from the system.
2. The desk is subjected to three *external* forces. They are caused by:
   i. (2) Reactions at the pin support **ii.**
   iii. Weight of bar
3. Draw the free-body diagram of the (detached) bar showing all these forces labeled with their magnitudes and directions. Include any other relevant information e.g. lengths, angles etc which may help when formulating the equations of motion (including the moment equation) for the bar.
4. Draw the corresponding kinetic diagram for the bar at the instant in question.

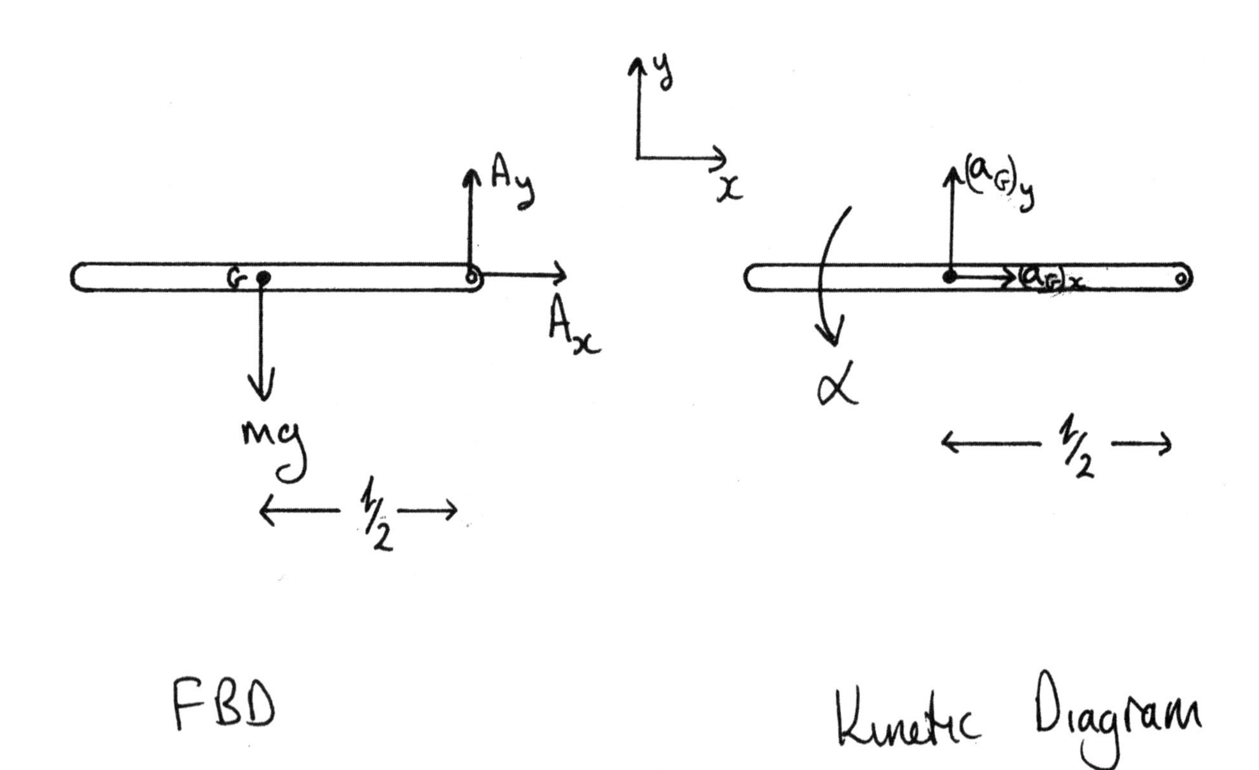

Problem 3.21

The gears A and B can turn freely on their pin supports in the vertical plane. Their masses and moments of inertia are m_A, m_B, I_A and I_B, respectively. They are initially stationary and , at $t = 0$, a constant couple of magnitude M is applied to gear B. Draw free-body and kinetic diagrams for each of the gears at this instant.

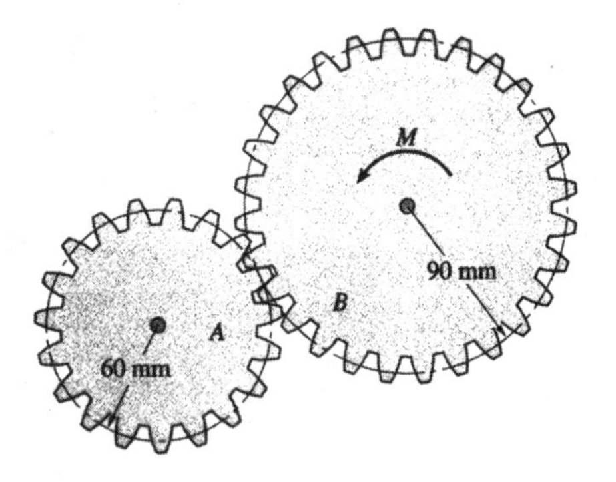

Solution

1. Imagine each gear to be separated or detached from the system.
2. Gear B is subjected to four *external* forces and one couple moment. They are caused by:

   i. **ii.**

   iii. **iv.**

   v.

   Gear A is subjected to four *external* forces. They are caused by:

   i. **ii.**

   iii. **iv.**
3. Draw the free-body diagram of each (detached) gear showing all these forces labeled with their magnitudes and directions. Include any other relevant information e.g. lengths, angles etc which may help when formulating the equations of motion (including the moment equation) for the gears.
4. Draw the corresponding kinetic diagrams for each gear.

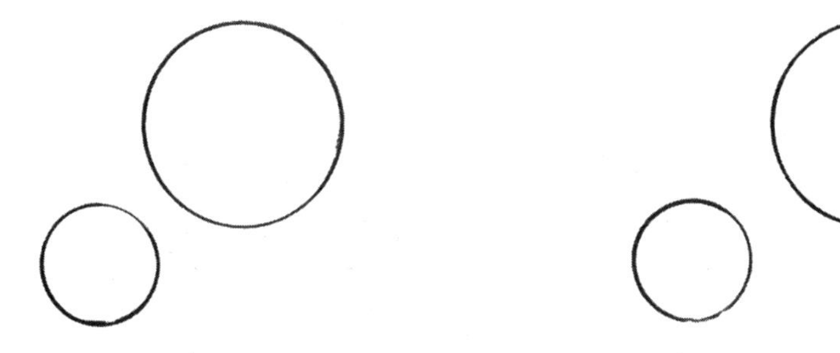

Problem 3.21

The gears A and B can turn freely on their pin supports in the vertical plane. Their masses and moments of inertia are m_A, m_B, I_A and I_B, respectively. They are initially stationary and , at $t = 0$, a constant couple of magnitude M is applied to gear B. Draw free-body and kinetic diagrams for each of the gears at this instant.

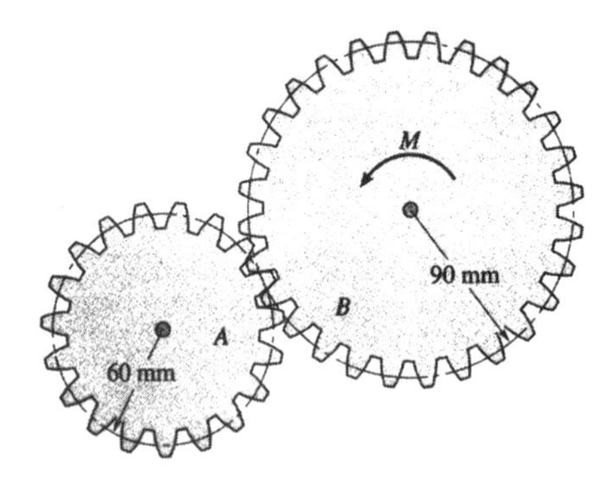

Solution

1. Imagine each gear to be separated or detached from the system.
2. Gear B is subjected to four *external* forces and one couple moment. They are caused by:
   - **i. Weight**
   - **ii. (2) Pin reactions**
   - **iii. Friction at contact with gear** A
   - **iv. Couple of magnitude** M

   Gear A is subjected to four *external* forces. They are caused by:
   - **i. Weight**
   - **ii. (2) Pin reactions**
   - **iii. Friction at contact with gear** B
3. Draw the free-body diagram of each (detached) gear showing all these forces labeled with their magnitudes and directions. Include any other relevant information e.g. lengths, angles etc which may help when formulating the equations of motion (including the moment equation) for the gears.
4. Draw the corresponding kinetic diagrams for each gear.

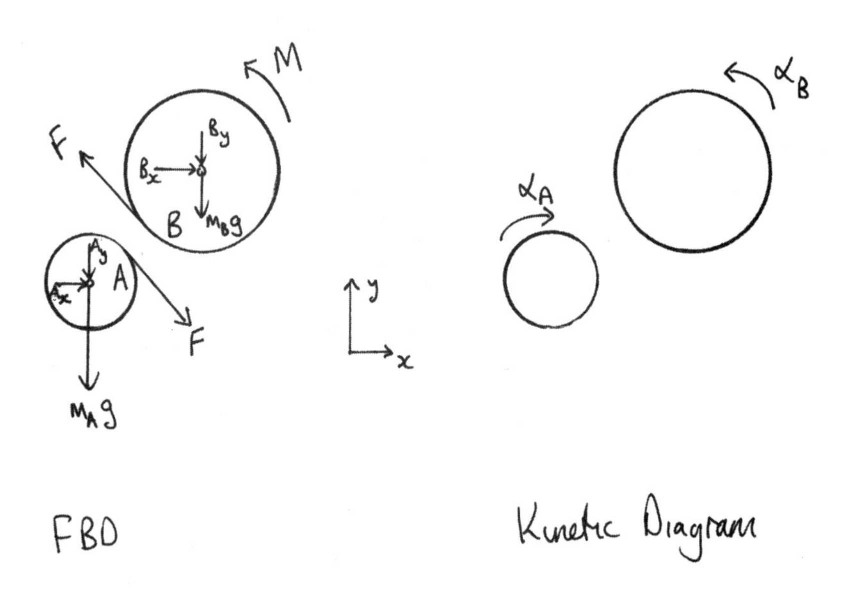

Problem 3.22

The moment of inertia of the pulley is I_P slug-ft^2. Draw separate free-body and kinetic diagrams for the pulley and the 20 lb weight.

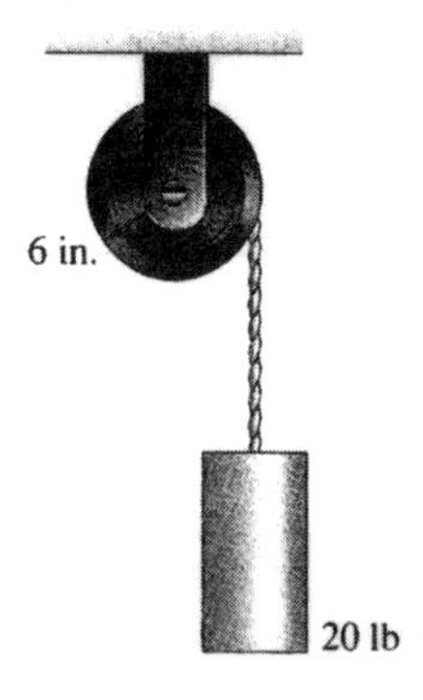

Solution

1. Imagine first the pulley then the weight to be separated or detached from the system.
2. The pulley is subjected to four *external* forces. They are caused by:

   i. **ii.**

   iii. **iv.**

   The weight is subjected to two *external* forces. They are caused by:

   i. **ii.**

3. Draw the free-body diagram of each of the (detached) pulley and weight showing all these forces labeled with their magnitudes and directions. Include any other relevant information e.g. lengths, angles etc which may help when formulating the equations of motion (including the moment equation) for the weight and/or pulley.
4. Draw the corresponding kinetic diagrams for each of the weight and pulley.

Problem 3.22

The moment of inertia of the pulley is I_P slug-ft^2. Draw separate free-body and kinetic diagrams for the pulley and the 20 lb weight.

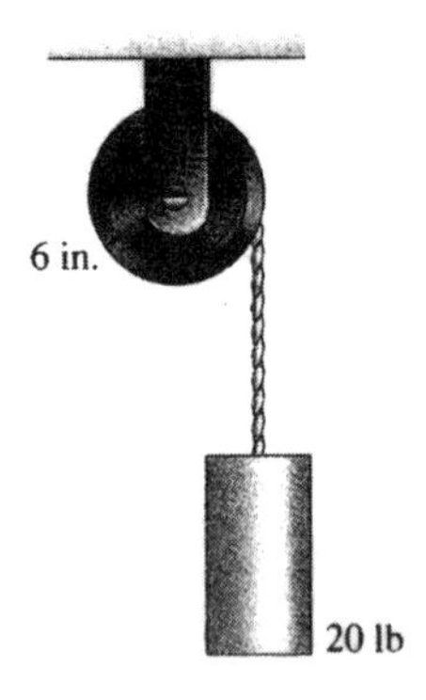

Solution

1. Imagine first the pulley then the weight to be separated or detached from the system.
2. The pulley is subjected to four *external* forces. They are caused by:
   - **i. Weight**
   - **ii. (2) Pin reactions**
   - **iii. Tension in rope**

   The weight is subjected to two *external* forces. They are caused by:
   - **i. Weight**
   - **ii. Tension in rope**
3. Draw the free-body diagram of each of the (detached) pulley and weight showing all these forces labeled with their magnitudes and directions. Include any other relevant information e.g. lengths, angles etc which may help when formulating the equations of motion (including the moment equation) for the weight and/or pulley.
4. Draw the corresponding kinetic diagrams for each of the weight and pulley.

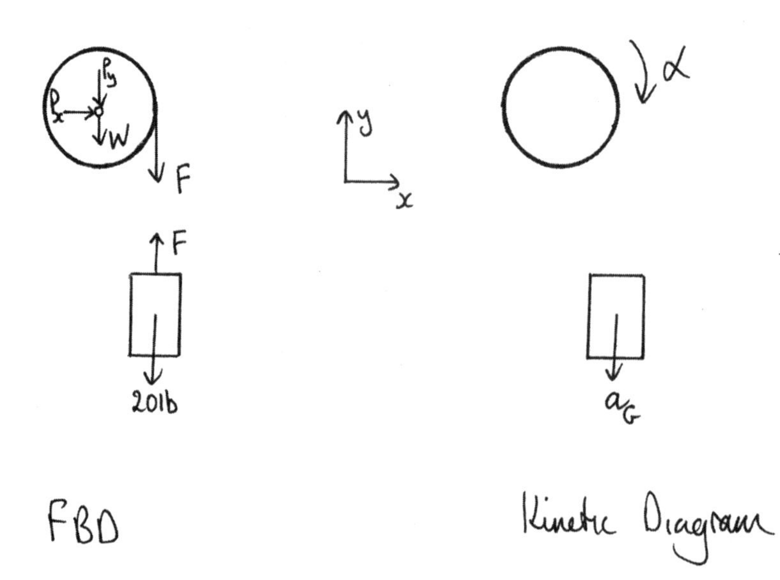

Problem 3.23

Box A has mass m_A, box B has mass m_B and the pulley has mass m_P. If the boxes start from rest at $t = 0$, draw free-body diagrams for each of the boxes and the pulley as box B slides down the slope. The coefficient of kinetic friction between the surface and the boxes is μ_k.

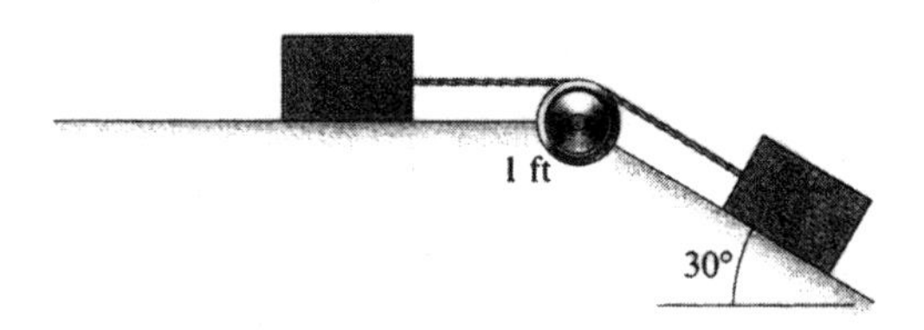

Solution

1. Imagine each box and the pulley to be separated or detached from the system.
2. Each box is subjected to four *external* forces. They are caused by:

   i. **ii.**
   iii. **iv.**

   The pulley is subjected to five *external* forces. They are caused by:

   i. **ii.**
   iii. **iv.**
   v.
3. Draw the free-body diagrams of the (detached) boxes and pulley showing all these forces labeled with their magnitudes and directions. Include any other relevant information e.g. lengths, angles etc which may help when formulating the equations of motion (including the moment equation) for the boxes and/or the pulley.
4. Include any useful kinetic information on the free-body diagrams.

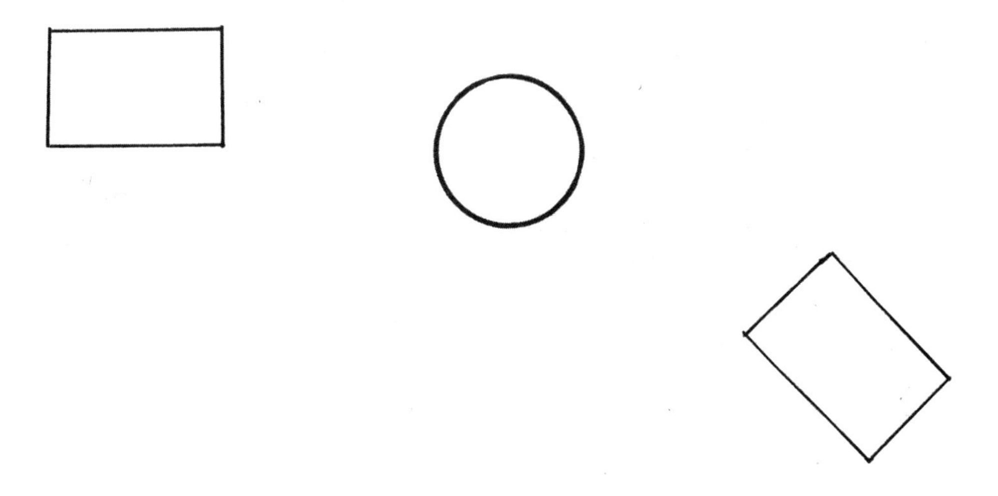

Problem 3.23

Box A has mass m_A, box B has mass m_B and the pulley has mass m_P. If the boxes start from rest at $t = 0$, draw free-body diagrams for each of the boxes and the pulley as box B slides down the slope. The coefficient of kinetic friction between the surface and the boxes is μ_k.

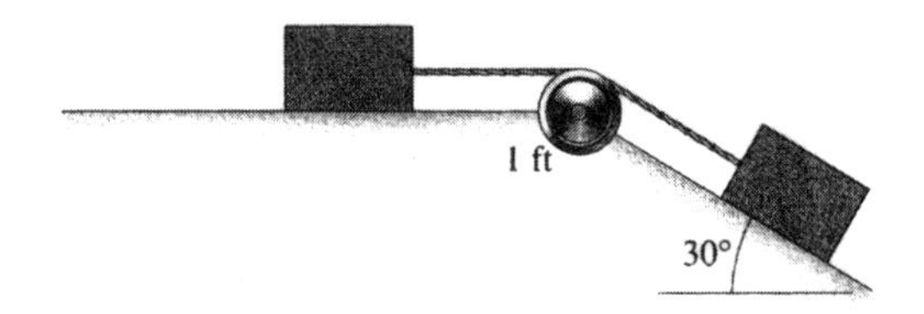

Solution

1. Imagine each box and the pulley to be separated or detached from the system.
2. Each box is subjected to four *external* forces. They are caused by:
   - **i. Weight**
   - **ii. Reaction at surface**
   - **iii. Friction at surface**
   - **iv. Tension in rope**

   The pulley is subjected to five *external* forces. They are caused by:
   - **i. Weight**
   - **ii. (2) pin reactions**
   - **iii. Tension in lower rope**
   - **iv. Tension in upper rope**
3. Draw the free-body diagrams of the (detached) boxes and pulley showing all these forces labeled with their magnitudes and directions. Include any other relevant information e.g. lengths, angles etc which may help when formulating the equations of motion (including the moment equation) for the boxes and/or the pulley.
4. Include any useful kinetic information on the free-body diagrams.

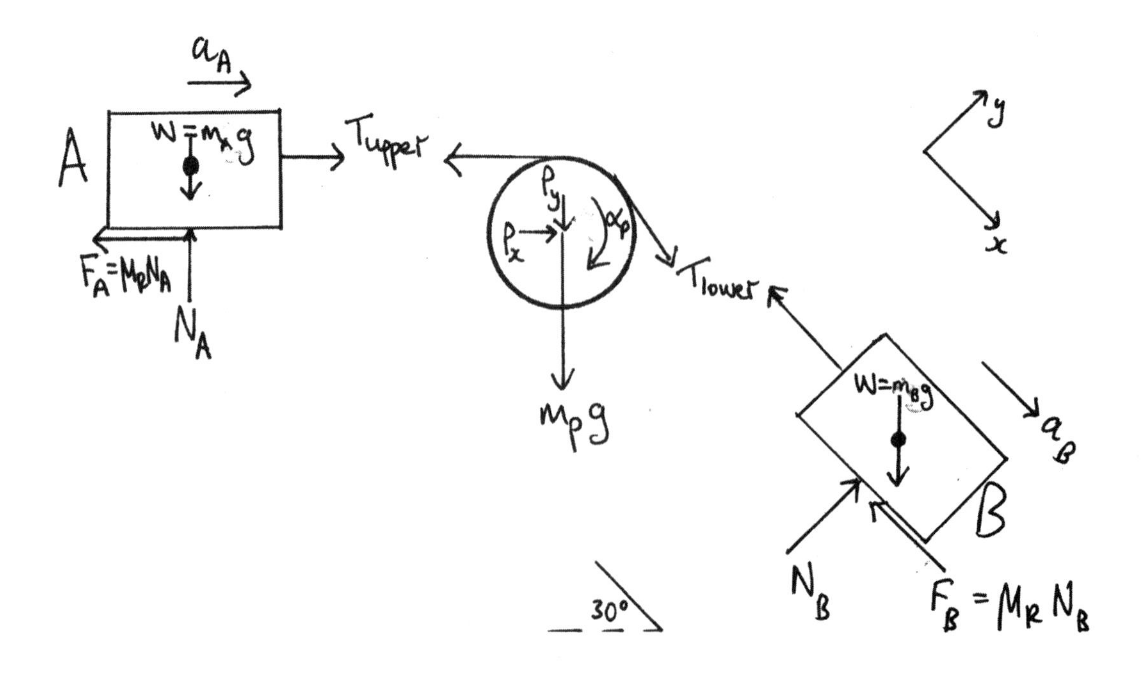

Problem 3.24

The slender bar weighs 10 lb and the disk weighs 20 lb. The coefficient of kinetic friction between the disk and the horizontal surface is μ_k. The disk has an initial counterclockwise angular velocity ω. Draw the free-body diagrams of the bar and the disk as the disk slides.

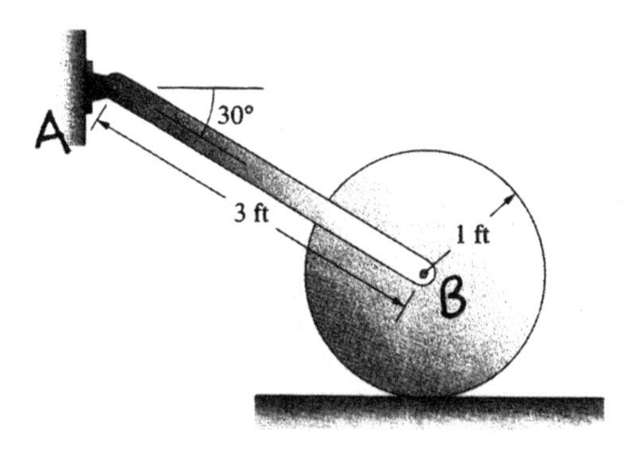

Solution

1. Imagine the disk and the bar each to be separated or detached from the system.
2. The disk is subjected to five *external* forces. They are caused by:

   i. **ii.**

   iii. **iv.**

   v.

   The bar is subjected to five *external* forces. They are caused by:

   i. **ii.**

   iii. **iv.**

   v.
3. Draw the free-body diagrams of the (detached) bar and disk showing all these forces labeled with their magnitudes and directions. Include any other relevant information e.g. lengths, angles etc which may help when formulating the equations of motion (including the moment equation) for the disk.
4. On the free-body diagrams, indicate clearly the acceleration components of the bar and disk.

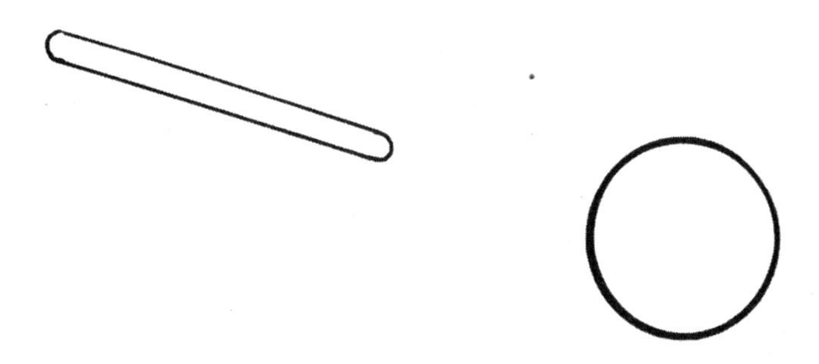

Problem 3.24

The slender bar weighs 10 lb and the disk weighs 20 lb. The coefficient of kinetic friction between the disk and the horizontal surface is μ_k. The disk has an initial counterclockwise angular velocity ω. Draw the free-body diagrams of the bar and the disk as the disk slides.

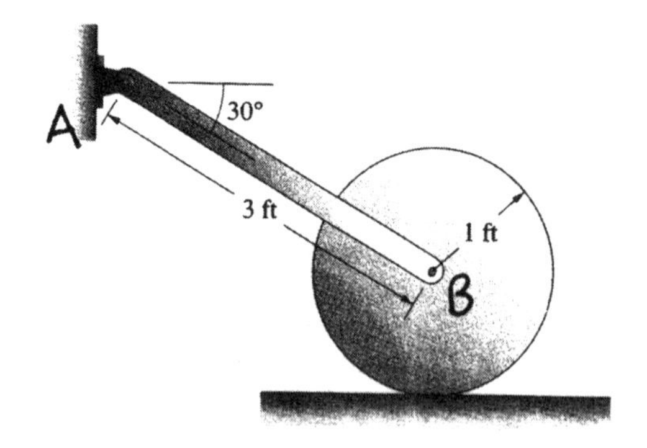

Solution

1. Imagine the disk and the bar each to be separated or detached from the system.
2. The disk is subjected to five *external* forces. They are caused by:
   - **i. Weight**
   - **ii. (2) pin reactions at** ***B***
   - **iii. Reaction at surface**
   - **iv. Friction at surface**

   The bar is subjected to five *external* forces. They are caused by:
   - **i. Weight**
   - **ii. (2) pin reactions at** ***A***
   - **iii. (2) pin reactions at** ***B***
3. Draw the free-body diagrams of the (detached) bar and disk showing all these forces labeled with their magnitudes and directions. Include any other relevant information e.g. lengths, angles etc which may help when formulating the equations of motion (including the moment equation) for the disk.
4. On the free-body diagrams, indicate clearly the acceleration components of the bar and disk.

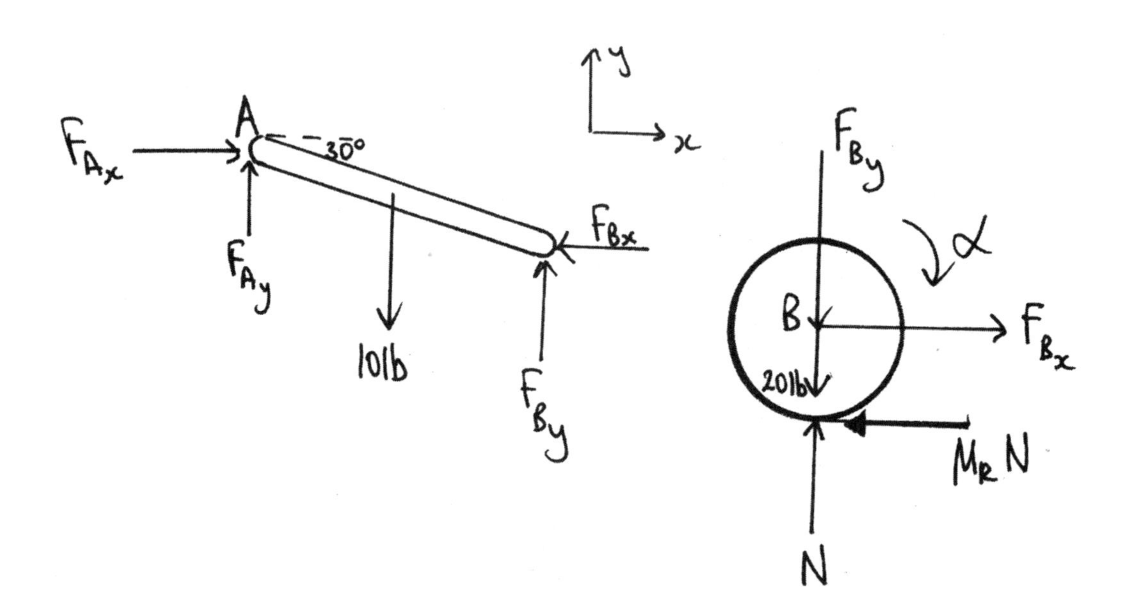

Problem 3.25

The object consists of identical 1–m, 5–kg bars welded together. If it is released from rest in the position shown, draw free-body and kinetic diagrams of the object at that instant. The center of mass of the system is shown.

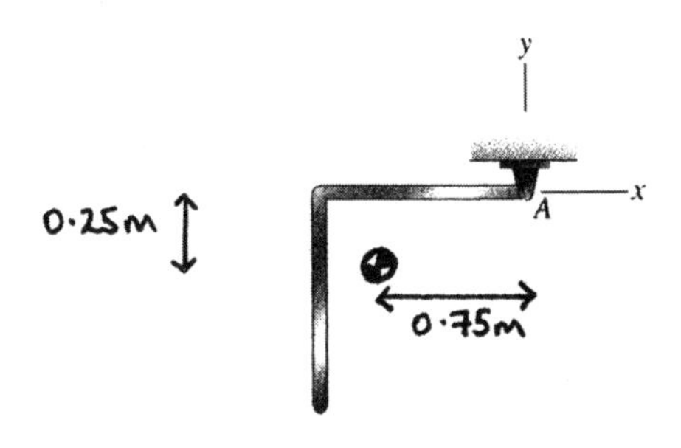

Solution

1. Imagine the object to be separated or detached from the pin at A.
2. The object is subjected to three *external* forces. They are caused by:

   i. **ii.**

   iii.
3. Draw the free-body diagram of the (detached) object showing all these forces labeled with their magnitudes and directions. Include any other relevant information e.g. lengths, angles etc which may help when formulating the equations of motion (including the moment equation) for the object.
4. Draw the corresponding kinetic diagram indicating clearly the acceleration components of the object.

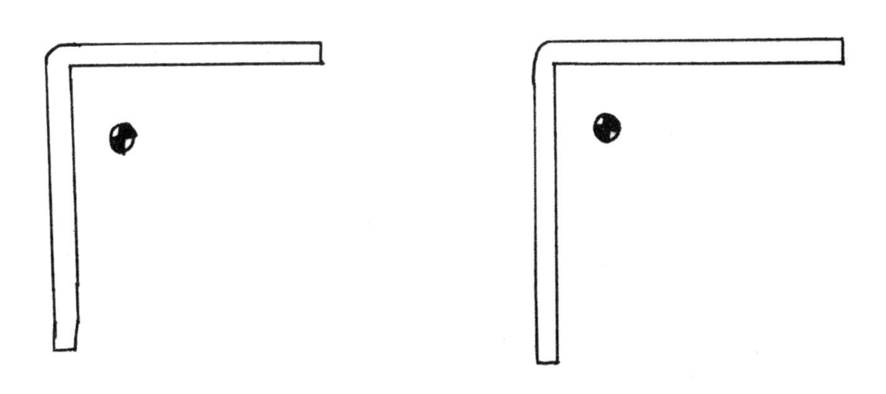

Problem 3.25

The object consists of identical 1−m, 5−kg bars welded together. If it is released from rest in the position shown, draw free-body and kinetic diagrams of the object at that instant. The center of mass of the system is shown.

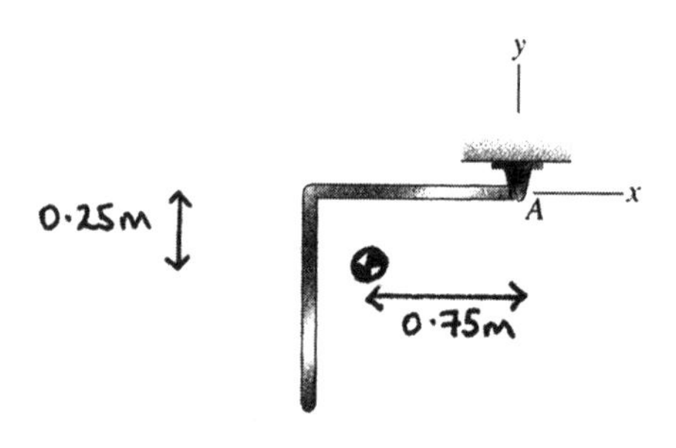

Solution

1. Imagine the object to be separated or detached from the pin at A.
2. The object is subjected to three *external* forces. They are caused by:

   i. Weight **ii. (2)pin reactions at** A

3. Draw the free-body diagram of the (detached) object showing all these forces labeled with their magnitudes and directions. Include any other relevant information e.g. lengths, angles etc which may help when formulating the equations of motion (including the moment equation) for the object.
4. Draw the corresponding kinetic diagram indicating clearly the acceleration components of the object.

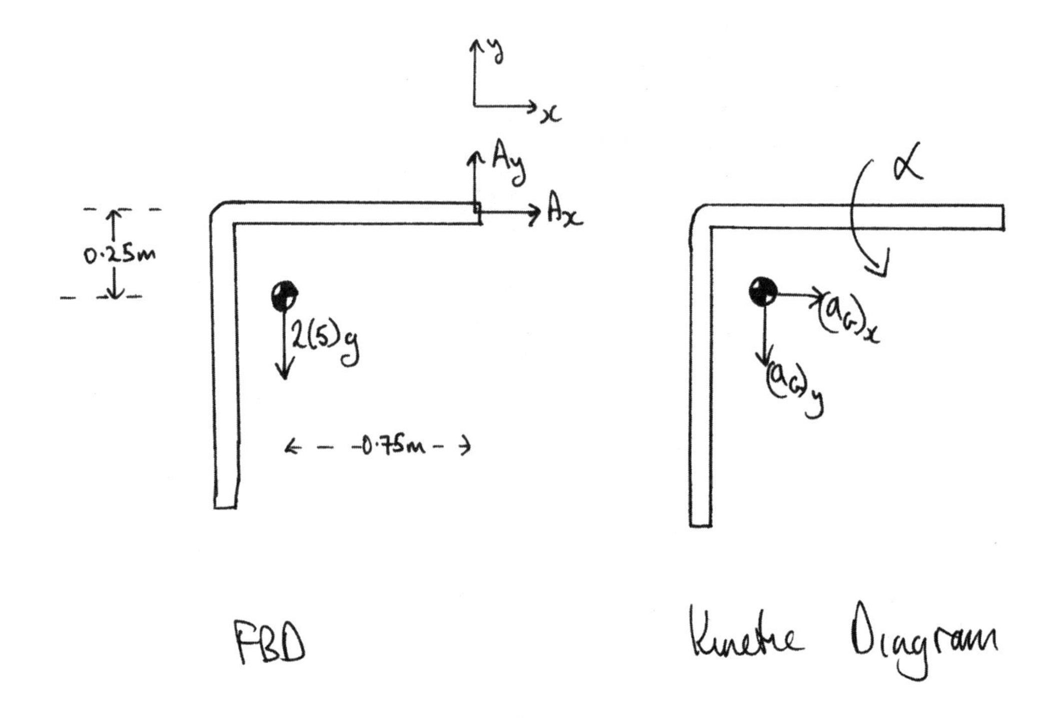

Problem 3.26

Model the arm ABC as a single rigid body. Its mass is 300 kg and the moment of inertia about its center of mass is I. If point A is stationary and the angular acceleration of the arm is $\boldsymbol{\alpha}_{arm}$, counterclockwise, draw a free-body diagram for the arm ABC.

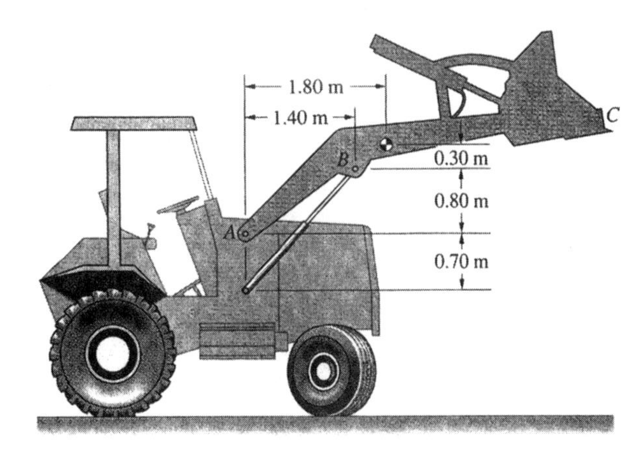

Solution

1. Imagine the arm ABC to be separated or detached from the system.
2. The arm is subjected to four *external* forces They are caused by:

   i. **ii.**

   iii. **iv.**

3. Draw the free-body diagram of the (detached) arm showing all these forces labeled with their magnitudes and directions. Include any other relevant information e.g. lengths, angles etc which may help when formulating the equations of motion (including the moment equation) for the arm.
4. Indicate the acceleration components of the arm on the coordinate axes system chosen in the free-body diagram.

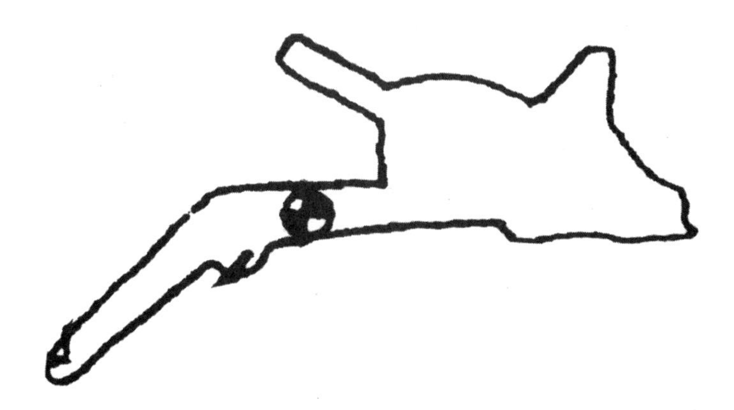

Problem 3.26

Model the arm ABC as a single rigid body. Its mass is 300 kg and the moment of inertia about its center of mass is I. If point A is stationary and the angular acceleration of the arm is $\boldsymbol{\alpha}_{arm}$, counterclockwise, draw a free-body diagram for the arm ABC.

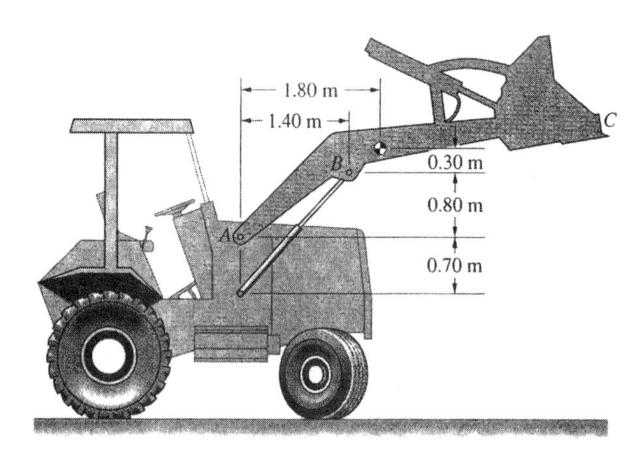

Solution

1. Imagine the arm ABC to be separated or detached from the system.
2. The arm is subjected to four *external* forces They are caused by:
   i. Weight **ii. (2) pin reactions at** A
   iii. Force exerted by hydraulic cylinder at B
3. Draw the free-body diagram of the (detached) arm showing all these forces labeled with their magnitudes and directions. Include any other relevant information e.g. lengths, angles etc which may help when formulating the equations of motion (including the moment equation) for the arm.
4. Indicate the acceleration components of the arm on the coordinate axes system chosen in the free-body diagram.

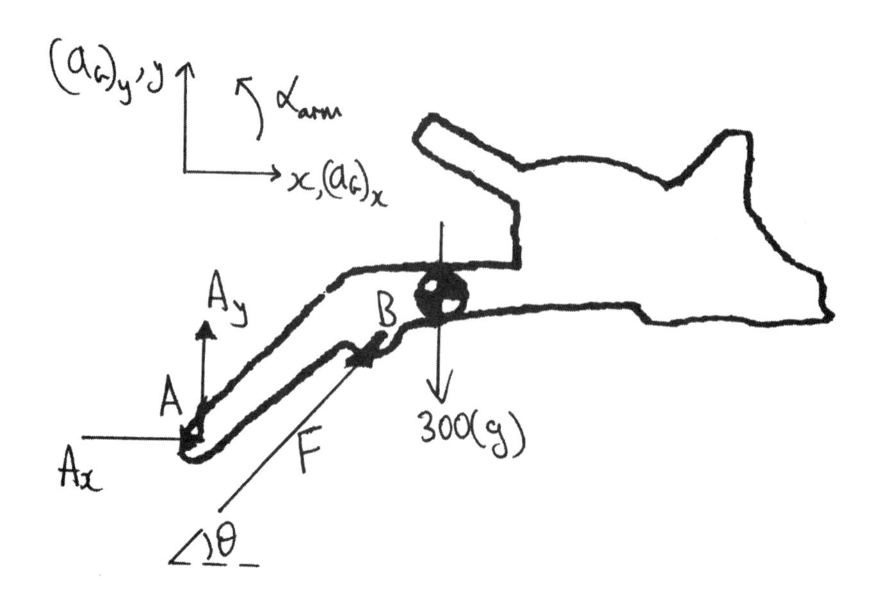

Problem 3.27

Arm BC has a mass of 12 kg and the moment of inertia about its center of mass is 3 kg-m^2. If B is stationary (fixed support) and arm BC has a constant counterclockwise angular velocity of 2 rad/s at the instant shown, use a free-body diagram of arm BC to determine the couple and the components of force exerted on arm BC at B.

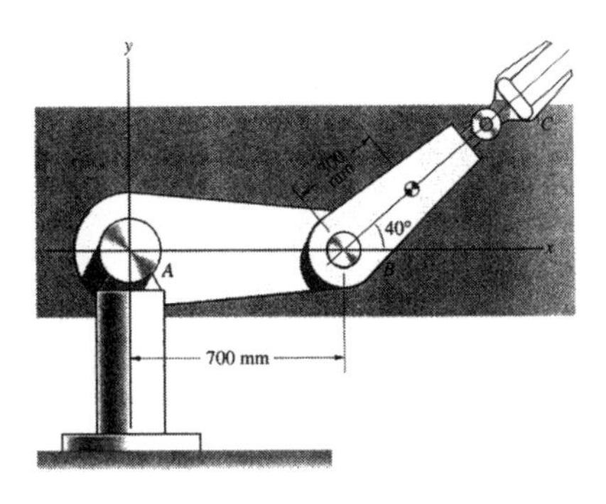

Solution

1. Imagine the arm to be separated or detached from the system.
2. The arm is subjected to three *external* forces and one external couple. They are caused by:

   i.   **ii.**

   iii.   **iv.**

   v.   **vi.**

3. Draw the free-body diagram of the (detached) arm showing all these forces labeled with their magnitudes and directions. Include any other relevant information e.g. lengths, angles etc which may help when formulating the equations of motion (including the moment equation) for the arm.
4. Indicate the acceleration components of the arm on the coordinate axes system chosen in the free-body diagram.

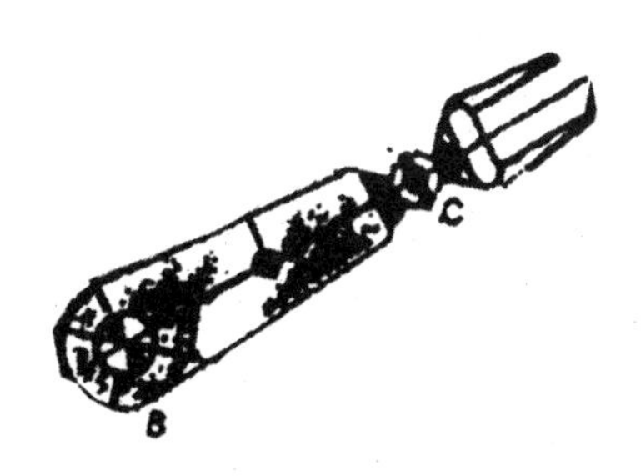

5. Sum moments about the fixed point B (the arm is in rotation about B) and write down the moment equation of motion:

   $\circlearrowleft + \sum M_B = I_B\alpha$:

6. Solve for the couple exerted at B.
7. Write down equations of motion in each of the $x-$ (horizontal) and $y-$ (vertical) directions:

   $+\rightarrow \; \sum F_x = m\,(a_G)_x$:

   $+\uparrow \; \sum F_y = m\,(a_G)_y$:

8. Use kinematics to determine the components $(a_G)_x$ and $(a_G)_y$:

9. Solve the equations of motion in the x and y directions to determine the reactions at B:

Problem 3.27

Arm BC has a mass of 12 kg and the moment of inertia about its center of mass is 3 kg-m^2. If B is stationary (fixed support) and arm BC has a constant counterclockwise angular velocity of 2 rad/s at the instant shown, use a free-body diagram of arm BC to determine the couple and the components of force exerted on arm BC at B.

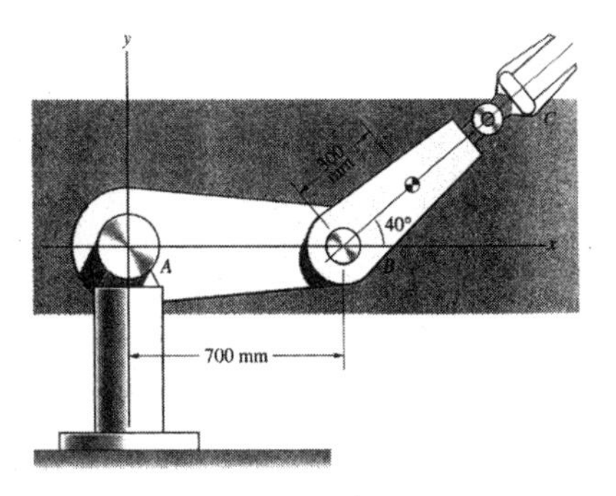

Solution

1. Imagine the arm to be separated or detached from the system.
2. The arm is subjected to three *external* forces and one external couple.
   They are caused by:
   i. (1) couple and (2) force reactions at B **ii. Weight**
3. Draw the free-body diagram of the (detached) arm showing all these forces labeled with their magnitudes and directions. Include any other relevant information e.g. lengths, angles etc. which may help when formulating the equations of motion (including the moment equation) for the arm.
4. Indicate the acceleration components of the arm on the coordinate axes system chosen in the free-body diagram.

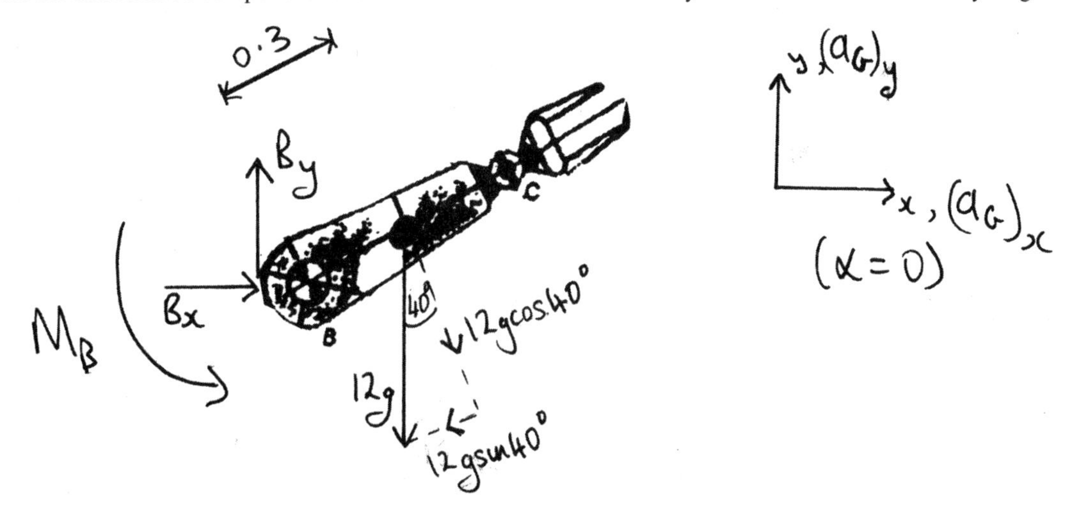

5. Sum moments about the fixed point B (the arm is in rotation about B) and write down the moment equation of motion:
   $\circlearrowleft + \sum M_B = I_B\alpha : M_B - 0.3\cos 40°(12)(9.81) = I_B\,(0)$
6. Solve for the couple exerted at B. $\mathbf{M}_B = 27.05\,\mathbf{k}$.
7. Write down equations of motion in each of the $x-$ (horizontal) and $y-$ (vertical) directions:
   $+\rightarrow \sum F_x = m\,(a_G)_x : B_x = m(a_G)_x = 12(a_G)_x$

   $+\uparrow \sum F_y = m\,(a_G)_y : B_y - mg = m\,(a_G)_y \Leftrightarrow B_y - 12g = 12(a_G)_y$

8. Use kinematics to determine the components $(a_G)_x$ and $(a_G)_y$:

$$\begin{aligned} \boldsymbol{a} &= (a_G)_x\mathbf{i} + (a_G)_x\mathbf{j} \\ &= \boldsymbol{\alpha} \times \mathbf{r}_{CM/O} - \omega^2\mathbf{r}_{CM/O} \\ &= -(2)^2(0.3\cos 40°\mathbf{i} + 0.3\sin 40°\mathbf{j} \\ &= -0.919\mathbf{i} - 0.771\mathbf{j}(\text{m/s}^2) \end{aligned}$$

$$(a_G)_x = -0.919, \quad (a_G)_y = -0.771$$

9. Solve the equations of motion in the x and y directions to determine the reactions at B:

$Bx = -11.03\ N,\ By = 108.5\ N.$

Problem 3.28

A circular disk of mass m and radius R is released from rest on an inclined surface and allowed to roll a distance D. Draw free-body and kinetic diagrams for the disk as it rolls.

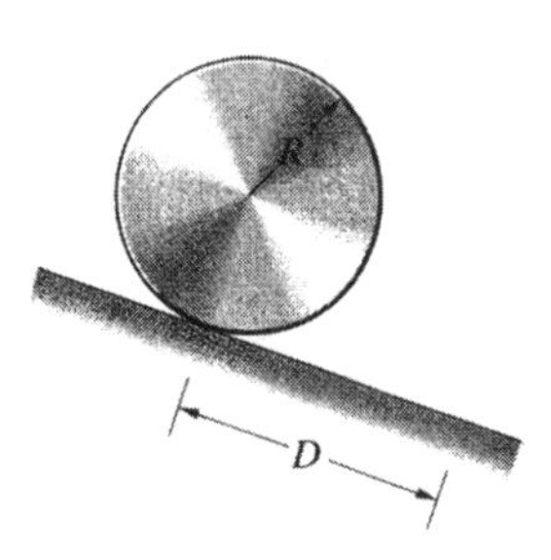

Solution

1. Imagine the disk to be separated or detached from the system.
2. The disk is subjected to three *external* forces. They are caused by:

   i. **ii.**

   iii.

3. Draw the free-body diagram of the (detached) disk showing all these forces labeled with their magnitudes and directions. Include any other relevant information e.g. lengths, angles etc which may help when formulating the equations of motion (including the moment equation) for the disk.
4. Indicate the acceleration components of the disk on the corresponding kinetic diagram.

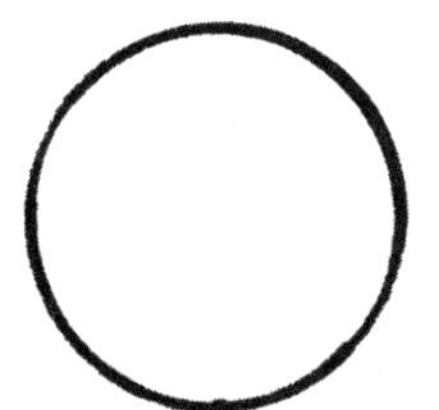

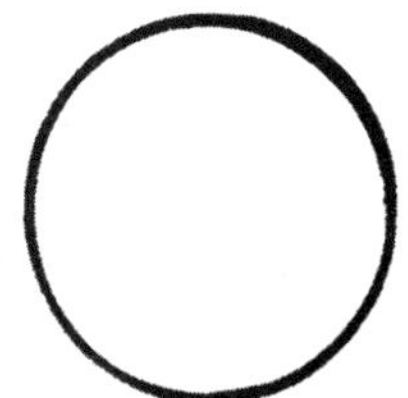

Problem 3.28

A circular disk of mass m and radius R is released from rest on an inclined surface and allowed to roll a distance D. Draw free-body and kinetic diagrams for the disk as it rolls.

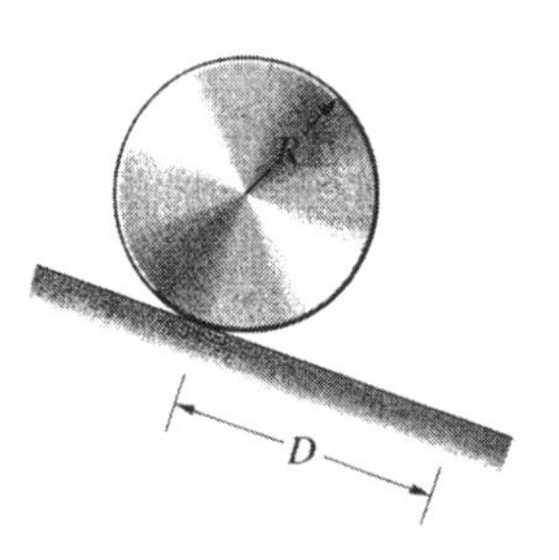

Solution

1. Imagine the disk to be separated or detached from the system.
2. The disk is subjected to three external forces. They are caused by:
   - **i. Weight**
   - **ii. Normal reaction at surface**
   - **iii. Friction at surface**
3. Draw the free-body diagram of the (detached) disk showing all these forces labeled with their magnitudes and directions. Include any other relevant information e.g. lengths, angles etc which may help when formulating the equations of motion (including the moment equation) for the disk.
4. Indicate the acceleration components of the disk on the corresponding kinetic diagram.

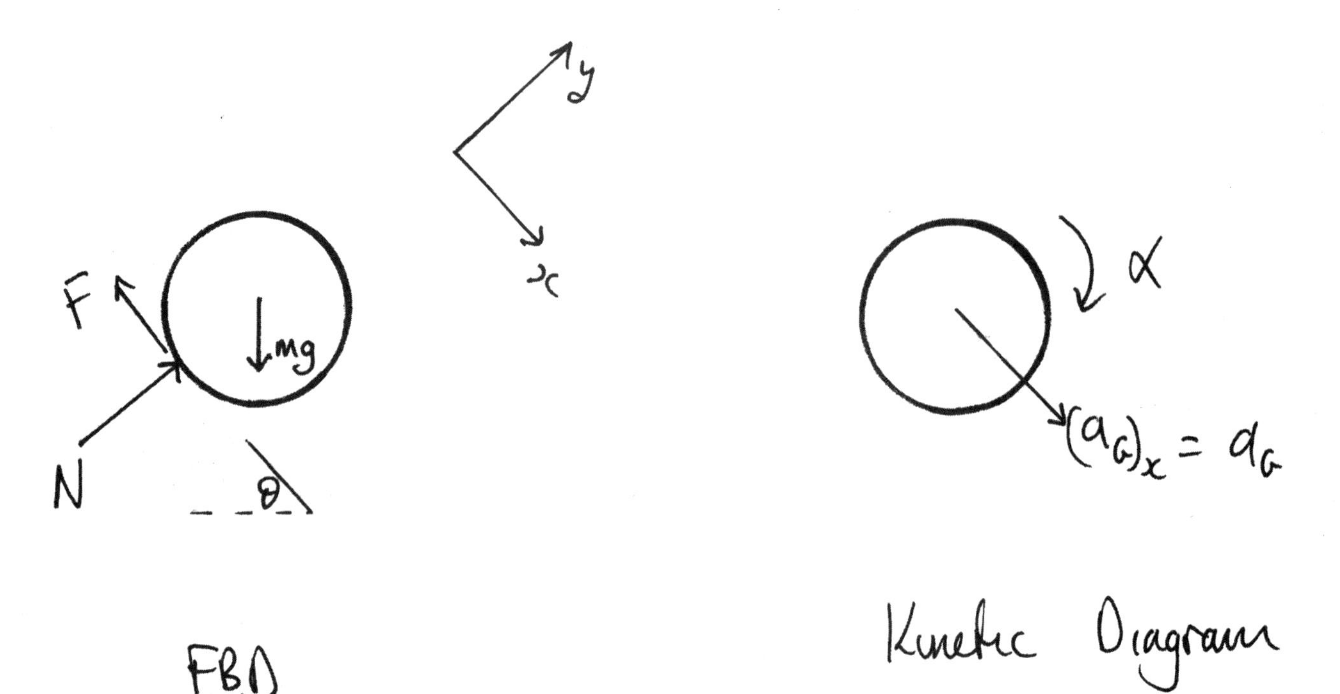

Problem 3.29

The homogeneous disk weighs 100 lb and its radius is $R = 1$ ft. It rolls on the plane surface. The spring constant is $k = 100$ lb/ft. If the disk is rolled to the left until the spring is compressed 1 ft and released from rest, use a free-body diagram of the disk to determine the disk's angular acceleration at the instant the disk is released.

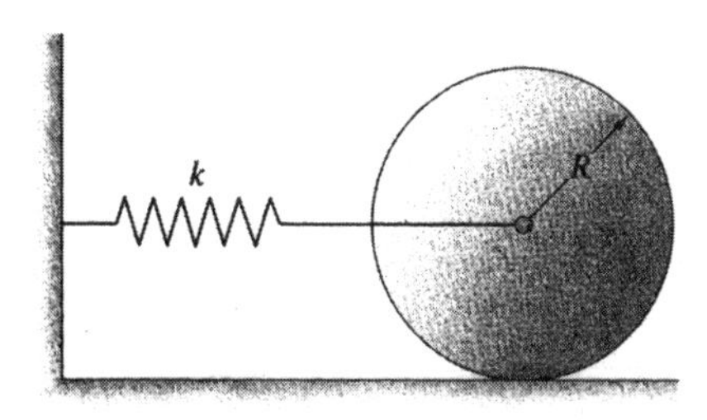

Solution

1. Imagine the disk to be separated or detached from the system.
2. The spool is subjected to four *external* forces. They are caused by:

   i. **ii.**

   iii. **iv.**

3. Draw the free-body diagram of the (detached) disk showing all these forces labeled with their magnitudes and directions. Include any other relevant information e.g. lengths, angles etc which may help when formulating the equations of motion (including the moment equation) for the spool.

4. Sum moments about the center of the disk and write down the moment equation of motion: $\circlearrowleft + \sum M_G = I_G \alpha$:
5. Write down equations of motion in each of the $x-$ (horizontal) and $y-$ (vertical) directions:

   $+\!\rightarrow \sum F_x = m\,(a_G)_x$:

   $+\!\uparrow \; \sum F_y = m\,(a_G)_y$:

6. Solve for the required angular acceleration $\boldsymbol{\alpha}$.

Problem 3.29

The homogeneous disk weighs 100 lb and its radius is $R = 1$ ft. It rolls on the plane surface. The spring constant is $k = 100$ lb/ft. If the disk is rolled to the left until the spring is compressed 1 ft and released from rest, use a free-body diagram of the disk to determine the disk's angular acceleration at the instant the disk is released.

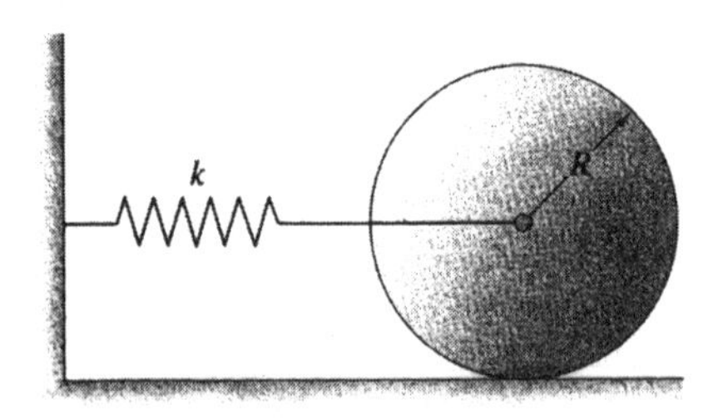

Solution

1. Imagine the disk to be separated or detached from the system.
2. The spool is subjected to four external forces. They are caused by:
   i. Weight
   ii. Spring force
   iii. Normal reaction at surface
   iv. Friction at surface
3. Draw the free-body diagram of the (detached) disk showing all these forces labeled with their magnitudes and directions. Include any other relevant information e.g. lengths, angles etc which may help when formulating the equations of motion (including the moment equation) for the spool.

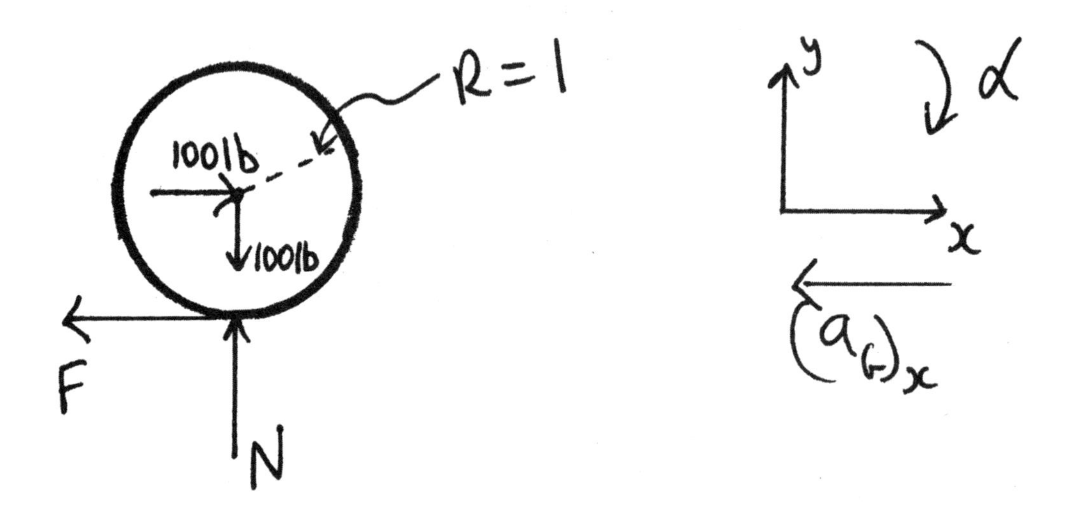

4. Sum moments about the center of the disk and write down the moment equation of motion:
   $\circlearrowleft + \sum M_G = I_G\alpha : -(1)(F) = I_G\alpha = \frac{1}{2}m\,(1)^2\,\alpha = \frac{1}{2}\left(\frac{100}{32.2}\right)(1)^2\alpha$
5. Write down equations of motion in each of the $x-$ (horizontal) and $y-$ (vertical) directions:
   $+\!\rightarrow \sum Fx = m\,(a_G)\,x : 100 - F = m\,(a_G)\,x = \left(\frac{100}{32.2}\right)(1)\alpha$ (disk rolls)

   $+\!\uparrow\ \sum Fy = m\,(a_G)\,y : N - 100 = 0$
6. Solve for the required angular acceleration $\boldsymbol{\alpha}$. $\boldsymbol{\alpha} = -21.5\,\mathbf{k}$

Problem 3.30

The ring gear is fixed. The mass and moment of inertia of the sun gear are m_s and I_s, respectively. The mass and moment of inertia of each planet gear are m_p and I_p, respectively. If a couple of magnitude M is applied to the sun gear, draw free-body and kinetic diagrams for the sun gear and planet gear A.

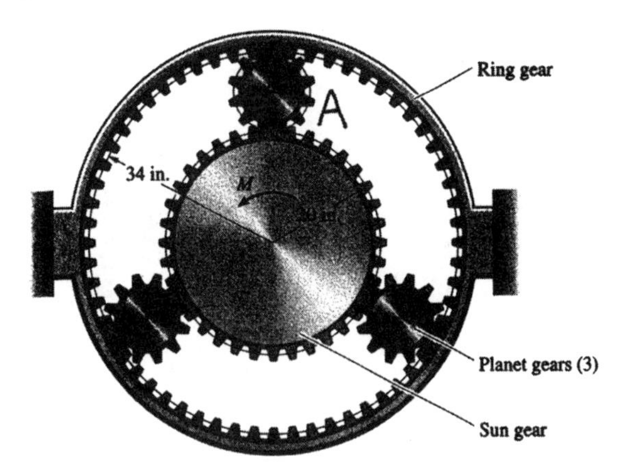

Solution

1. Imagine each of the sun and planet gear to be separated or detached from the system.
2. The sun gear is subjected to seven *external* forces and one external couple. The planet gear is subjected to five external forces.
3. Draw the free-body diagrams of the (detached) sun and planet gears showing all these forces labeled with their magnitudes and directions. Include any other relevant information e.g. lengths, angles etc which may help when formulating the equations of motion (including the moment equation) for the gears.
4. Draw the corresponding kinetic diagrams indicating clearly the acceleration components of each gear.

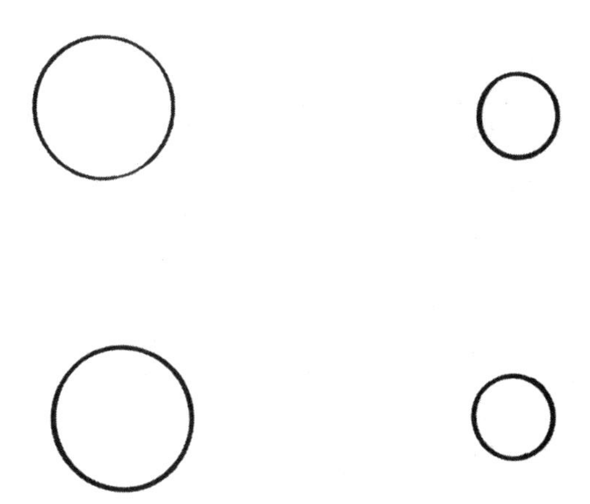

Problem 3.30

The ring gear is fixed. The mass and moment of inertia of the sun gear are m_s and I_s, respectively. The mass and moment of inertia of each planet gear are m_p and I_p, respectively. If a couple of magnitude M is applied to the sun gear, draw free-body and kinetic diagrams for the sun gear and planet gear A.

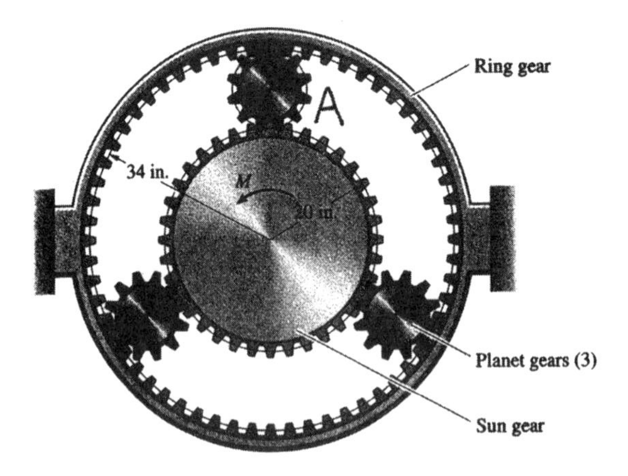

Solution

1. Imagine each of the sun and planet gear to be separated or detached from the system.
2. The sun gear is subjected to seven external forces and one external couple. The planet gear is subjected to five external forces.
3. Draw the free-body diagrams of the (detached) sun and planet gears showing all these forces labeled with their magnitudes and directions. Include any other relevant information e.g. lengths, angles etc which may help when formulating the equations of motion (including the moment equation) for the gears.
4. Draw the corresponding kinetic diagrams indicating clearly the acceleration components of each gear.

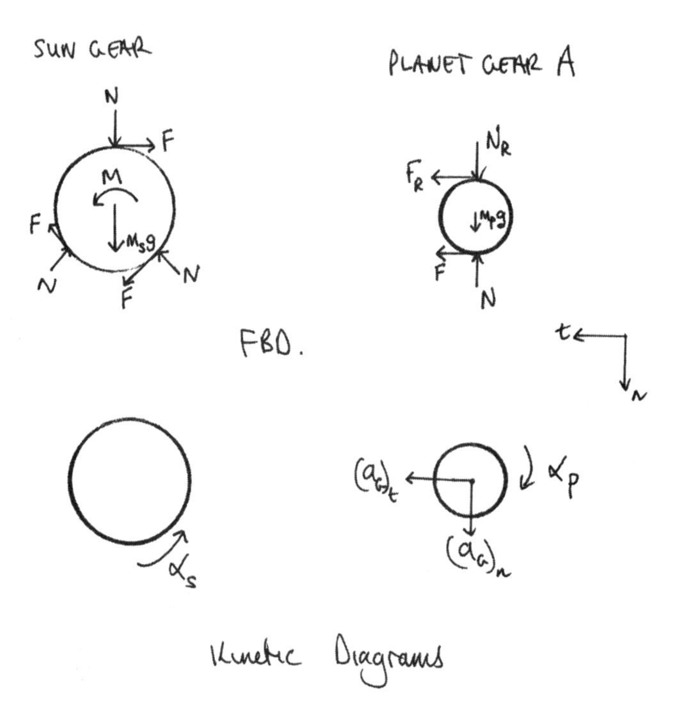

PART II

SECTION-BY-SECTION, CHAPTER-BY-CHAPTER SUMMARIES WITH REVIEW QUESTIONS AND ANSWERS.

12

Introduction

Main Goals of this Chapter:

- To introduce the basic ideas of *Mechanics*.
- To outline a general procedure for solving problems in mechanics.
- To review fundamental concepts in mechanics
- To review the principles for applying the SI and U.S. Customary systems of units.
- To give a concise statement of Newton's theory of gravitation.

12.1 Engineering and Mechanics

Modern engineers use *mathematical models* to predict and analyze the behavior of complex engineering systems. As an analytical science, *mechanics* is one of the main sources of fundamental concepts and analytical methods used in the design and analysis of mathematical models and hence engineering systems.

Mechanics is that branch of the physical sciences concerned with the behavior of bodies subjected to the action of forces. The subject of mechanics is divided into two parts:

- *statics*—the study of objects in equilibrium (objects either at rest or moving with a constant velocity).
- *dynamics*—the study of objects with accelerated motion.

  Although statics can be considered as a special case of dynamics (in which the acceleration is zero), it deserves special treatment since many objects are designed with the intention that they remain in equilibrium.

  The subject of *dynamics* is often itself divided into two parts:

  — *kinematics*—treats only the *geometric aspects* of the motion.

  — *kinetics*—analysis of *forces* causing the motion.

12.2 Learning Mechanics

- The most effective way to learn engineering mechanics is to *solve problems*. The following problem-solving procedure will be useful:
- *Problem Solving*

— *Identify* the information that is given and the information, or answer, you must determine. It's often helpful to restate the problem in your own words. When appropriate, make sure you understand the physical system or model involved.

— Develop a *strategy* for the problem. This means identifying the principles and equations that apply and deciding how you will use them to solve the problem. Whenever possible, draw diagrams to help visualize and solve the problem.

— Whenever you can, try to *predict* the answer. This will develop your intuition and will often help you recognize an incorrect answer.

— *Solve* the equations, and, whenever possible, interpret your results and compare them with your prediction. This last step is a *reality check*. Is your answer reasonable?

12.3 Fundamental Concepts

Numbers

- *Significant Digits*. This term refers to the number of meaningful (accurate) digits in a number, counting to the right starting with the first nonzero digit. The two numbers 7.630 and 0.007630 are each stated to four significant digits. If only the first four digits in the number 7,630,000 are known to be accurate, this can be indicated by writing the number in scientific notation as 7.630×10^6.

Space and Time

- *Space* simply refers to the three-dimensional universe in which we live. The distance between two points in space is the *length* of the straight line joining them. In *SI* units, the unit of length is the meter (m). In U.S. Customary units, the unit of length is the foot (ft).
- *Time* is measured by the intervals between repeatable events, such as the swings of a clock pendulum. In both *SI* and U.S. Customary units, the unit of time is the second (s).
- If the position of a point is space relative to some reference point changes with time, the rate of change of its position is called its *velocity*, and the rate of change of its velocity is called its *acceleration*. In *SI* units, the velocity is measured in *meters per second* (m/s) and the acceleration in *meters per second per second, or meters per second squared* (m/s^2). In *U.S. Customary* units, the velocity is expressed in *feet per second* (ft/s) and the acceleration is expressed in *feet per second squared* (ft/s^2).

Newton's Laws

Newton's laws apply to the motion of a particle (recall that a *particle* has a mass but a size that can be neglected i.e., the geometry of the body is ignored. A particle is often represented by a *point* in space.) as measured from a nonaccelerating (inertial) reference frame.

* **First Law**. *When the sum of the forces acting on a particle is zero, its velocity is constant. In particular, if the particle is initially stationary, it will remain stationary.*
* **Second Law**. *When the sum of the forces acting on a particle is non-zero, the sum of the forces is equal to the rate of change of the linear momentum of the particle. If the mass is constant, the sum of the forces is equal to the product of the mass of the particle and its acceleration.* i.e.

$$\mathbf{F} = m\mathbf{a}.$$

* **Third Law**. *The forces exerted by two particles on each other are equal in magnitude and opposite in direction.*

Note. Newton's second law gives precise meanings to the terms *mass* and *force*. In *SI* units, the unit of mass is the *kilogram* (kg). The unit of force is the *Newton* (N), which is the force required to give a mass of one kilogram an acceleration of one meter per second squared. In *U.S. Customary* units, the unit of force is the *pound* (lb). The unit of mass is the *slug*, which is the amount of mass accelerated at one foot per second squared by a force of one pound.

12.4 Units

The four basic quantities *force, mass, length*, and *time* are related by Newton's second law. Hence, the units used to define these quantities are not independent i.e., three of the four units are called *base units* (arbitrarily defined) and the fourth unit a *derived unit* (derived from Newton's second law).

International System of Units (SI Units)

- In the SI system, the unit of force, the *newton*, is a derived unit. The meter, second and kilogram are base units.
- One *newton* is equal to a force required to give one kilogram of mass an acceleration of 1 m/s^2.
- The common prefixes used in SI units are:

Prefix	Abbreviation	Multiple
nano-	n	10^{-9}
micro-	μ	10^{-6}
milli-	m	10^{-3}
kilo-	k	10^{3}
mega-	M	10^{6}
giga-	G	10^{9}

U.S. Customary Units

- In the U.S. Customary system, the unit of mass, the *slug*, is a derived unit. The foot, second and pound are base units.
- One *slug* is equal to the amount of matter accelerated at 1 ft/s^2 when acted upon by a force of 1 lb.

The following table summarizes the two systems of units.

Name	Length	Time	Mass	Force
International System (SI)	*meter(m)*	*second(s)*	*kilogram(kg)*	*newton** $\left(N = \dfrac{kg \cdot m}{s^2}\right)$
U.S. Customary	*foot(ft)*	*second(s)*	*slug** $\left(= \dfrac{lb \cdot s^2}{ft}\right)$	*pound(lb)*
**Derived Unit*				

Angular Units

- In both *SI* and *U.S.* Customary units, angles are normally expressed in *radians*(*rad*). 360° equals 2π radians.
- Equations containing angles are nearly always derived under the assumption that angles are expressed in radians. Therefore, when you want to substitute the value of an angle expressed in degrees into an equation, *you should first convert it into radians.*

Conversion of Units

The following table provides a set of direct conversion factors between U.S. Customary and SI units for the basic quantities.

Quantity	Unit (U.S. Customary)	Equals	Unit (SI)
Force	*lb*		4.448 *N*
Mass	*slug*		14.59 *kg*
Length	*ft*		0.3048 *m*
Length	*in*		25.4 *mm*

Note also that in the U.S. Customary system

1 ft = 12 *in (inches).*
5280 ft = 1 *mi (mile).*
1000 *lb* = 1 *kip (kilo-pound).*
2000 *lb* = 1 *ton.*

12.5 Newtonian Gravitation

$$F = G\frac{m_1 m_2}{r^2}$$

F = force of gravitation between two particles

G = universal constant of gravitation

m_1, m_2 = mass of each of the two particles

r = distance between the two particles

Note. The *weight* of a body has magnitude

$$W = mg,$$

where g is the acceleration due to gravity at sea level. The value of g varies from location to location on the surface of the earth but here is taken to be $g = 9.81\ m/s^2$ in *SI* units and $g = 32.2\ ft/s^2$ in U.S. Customary units.

Helpful Tips and Suggestions

- The *language* of engineering mechanics is *mathematics*. Consequently, make sure you review/re-read the necessary mathematical notation/concepts *as they arise* in your mechanics course (trying to review all of the necessary mathematics *at once* is not recommended—there's just too much to digest at one time). You should aim to achieve *fluency* in basic mathematical techniques/notation so that your learning of mechanics is not distracted by trying to remember things which your instructor *assumes* you know e.g., how to solve linear systems of algebraic equations, how to perform basic vector algebra, differentiation and integration etc.
- *Remember* that in solving problems from engineering mechanics you are solving real practical problems and producing real data with physical significance. Thus, you are responsible for making sure your results are correct, consistent and well-presented. Get into the habit of doing this *now* so that it will become second nature by the time you graduate. In the world of professional engineering you have a responsibility to your profession and to the many people that will use the product you will help to design, manufacture or implement.

Review Questions: True or False[1]?

1. The subject called *Dynamics* studies only bodies which are at rest.
2. The best way to learn mechanics is to solve relevant problems.
3. The numbers 3.456 and 0.003456 are each stated to 4 significant digits.
4. Newton's second law can be written as $\mathbf{F} = m\mathbf{a}$ irrespective of whether mass is constant or varies with time.
5. In the *SI* system of units, the *newton* is a derived unit.
6. In the *U.S.* Customary system of units, the *pound* (*lb*) is the derived unit.
7. It's OK to mix degrees with radians in any equation involving angles.
8. Newtonian gravitation can be used to approximate the weight of an object of mass m due to the gravitational attraction of the earth.
9. Weight is a property of matter that does not change from one location to another.
10. If you know an object's mass, its weight at sea level can be determined by $W = mg$.

[1] 1. F 2. T 3. T 4. F 5. T 6. F 7. F 8. T 9. F 10. T

13

Motion of a Point

Main Goals of this Chapter:

- To describe and analyze the motion of a point relative to a given reference frame.
- To introduce the concepts of position, velocity, and acceleration of a point.
- To study motion of a point along a straight line.
- To investigate motion of a point along a curved path using different coordinate systems:
  - ♦ Cartesian coordinates.
  - ♦ Normal and tangential coordinates.
  - ♦ Polar coordinates.

13.1 Position, Velocity, and Acceleration

- The *position* of a point P relative to a given reference frame with origin O can be specified by the position vector $\mathbf{r}$ from O to P.
- The *velocity* of P relative to the reference frame is

$$\mathbf{v} = \frac{d\mathbf{r}}{dt}.$$

- The *acceleration* of P relative to the reference frame is

$$\mathbf{a} = \frac{d\mathbf{v}}{dt}.$$

- *A point has the same velocity and acceleration relative to any fixed point in a given reference frame.*

13.2 Straight Line Motion

Description of the Motion

- The *position* of a point P on a straight line relative to a reference point O is specified by a coordinate s measured along the line from O to P.

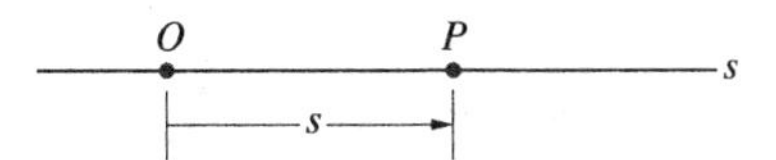

The coordinate s from O to P.

- The *velocity* of P along the line is related to the coordinate s by

$$v = \frac{ds}{dt},$$

where v is the magnitude of the velocity $\mathbf{v}$ and is commonly referred to as the *speed*.
- The *acceleration* of P along the line is related to the coordinate s by

$$a = \frac{dv}{dt},$$

where a is the magnitude of the acceleration $\mathbf{a}$.

Acceleration Specified as a Function of Time

If the acceleration is specified as a function of time, the velocity and position can be determined as functions of time by integration:

- **Velocity:**

$$\frac{dv}{dt} = a(t) \Longrightarrow v(t) = \int a\,(t)\,dt + A, \tag{13.1}$$

where A is an integration constant.
- **Position:**

$$\frac{ds}{dt} = v(t) \Longrightarrow s(t) = \int v(t)dt + B, \tag{13.2}$$

where B is an integration constant.

Note. The constants A and B can be determined when additional information is given about the motion, for example, the values of v and s at a given time.

Constant Acceleration

- Let $a = a_0 =$ constant. Assume that $v = v_0$ and $s = s_0$ at time $t = t_0$. Then, from Eqs. (12.1) and (12.2),

$$v = v_0 + a_0(t - t_0) \quad \text{(speed as a function of time)}, \tag{13.3}$$

$$s = s_0 + v_0(t - t_0) + \frac{1}{2}a_0(t - t_0)^2 \quad \text{(position as a function of time)}, \tag{13.4}$$

$$v^2 = v_0^2 + 2a_0\,(s - s_0) \quad \text{(speed as a function of position)}. \tag{13.5}$$

- *Eqs. (13.3)–(13.5) apply* ***only*** *when the acceleration is constant.*

Acceleration Specified as a Function of Velocity

If the acceleration is specified as a function of velocity, we can separate variables and integrate to obtain the velocity and position as functions of time:

- **Velocity.**

$$\frac{dv}{dt} = a(v)$$

$$\Longrightarrow \int_{v_0}^{v} \frac{dv}{a(v)} = \int_{t_0}^{t} dt \tag{13.6}$$

where v_0 is the magnitude of the velocity at time t_0..

- **Position.** In principle, we can solve Eq. (13.6) for v as a function of time and then integrate the relation

$$\frac{ds}{dt} = v$$

to determine the position as a function of time.
Note. We can also determine the *velocity as a function of position*. Writing the acceleration as

$$a(v) = \frac{dv}{dt} = \frac{dv}{ds}\frac{ds}{dt} = \frac{dv}{ds}v, \tag{13.7}$$

we have

$$\frac{vdv}{a(v)} = ds,$$

from which we obtain

$$\int_{v_0}^{v} \frac{vdv}{a(v)} = \int_{s_0}^{s} ds,$$

from which we obtain a relation between v and s.

Acceleration Specified as a Function of Position

- If the acceleration is specified as a function of position, $\frac{dv}{dt} = a(s)$, and, as in (13.7), the chain rule can be used to express the acceleration in terms of a derivative with respect to position:

$$\frac{dv}{dt} = \frac{dv}{ds}\frac{ds}{dt} = \frac{dv}{ds}v = a(s).$$

- **Velocity.** Separating variables, v can be determined as a function of position s

$$\int_{v_0}^{v} vdv = \int_{s_0}^{s} a(s)\, ds$$

$$\Longrightarrow v = \frac{ds}{dt} = v(s). \tag{13.8}$$

- **Position.** We can separate variables in (13.8) and integrate to determine the position as a function of time:

$$\int_{s_0}^{s} \frac{ds}{v(s)} = \int_{t_0}^{t} dt.$$

Summary

Table 13.1: Determining the velocity when you know the acceleration as a function of velocity or position

If you know $a = a(v)$,	separate variable: $\frac{dv}{dt} = a(v)$; $\frac{dv}{a(v)} = dt$. Or apply the chain rule $\frac{dv}{dt} = \frac{dv}{ds}\frac{ds}{dt} = \frac{dv}{ds}v = a(v)$, and then separate variables: $\frac{v\,dv}{a(v)} = ds$.
If you know $a = a(s)$,	apply the chain rule $\frac{dv}{dt} = \frac{dv}{ds}\frac{ds}{dt} = \frac{dv}{ds}v = a(s)$, and then separate variables: $v\,dv = a(s)ds$.

13.3 Curvilinear Motion

- *Curvilinear motion* occurs when the particle moves along a *curved path.*
- You have seen that the motion of a point along a straight line is described by the scalars s, v, and a. But, if a point describes a curvilinear path relative to some reference frame, we must specify its position in terms of position, velocity, and acceleration *vectors*.

Important Points

- Curvilinear motion can cause changes in *both* the magnitude and direction of the position, velocity, and acceleration vectors.
- The velocity vector is *always directed tangent* to the path.
- In general, the acceleration vector is *not* tangent to the path.

Cartesian Coordinates

Using a fixed xyz frame of reference:

- **Position.**

$$\mathbf{r}(t) = x(t)\mathbf{i} + y(t)\mathbf{j} + z(t)\mathbf{k},$$

$$\text{Magnitude} = r = \sqrt{x^2 + y^2 + z^2},$$

$$\text{Direction given by} = \frac{\mathbf{r}}{r}.$$

- **Velocity.**

$$\mathbf{v}(t) = \frac{d\mathbf{r}}{dt} = \frac{d}{dt}[x(t)\mathbf{i}] + \frac{d}{dt}[y(t)\mathbf{j}] + \frac{d}{dt}[z(t)]\,\mathbf{k},$$
$$= \frac{d}{dt}[x(t)]\,\mathbf{i} + \frac{d}{dt}[y(t)]\,\mathbf{j} + \frac{d}{dt}[z(t)]\,\mathbf{k}$$
$$= \frac{dx}{dt}\mathbf{i} + \frac{dy}{dt}\mathbf{j} + \frac{dz}{dt}\mathbf{k}$$
$$= v_x\mathbf{i} + v_y\mathbf{j} + v_z\mathbf{k}$$

$$\text{Magnitude} = v = \sqrt{v_x^2 + v_y^2 + v_z^2},\ \text{Direction: } \frac{\mathbf{v}}{v},\ \textit{always tangent to path.}$$

- **Acceleration.**

$$\mathbf{a}(t) = \frac{d\mathbf{v}}{dt} = a_x\mathbf{i} + a_y\mathbf{j} + a_z\mathbf{k}$$
$$\text{where } a_x = \dot{v}_x = \ddot{x},\ a_y = \dot{v}_y = \ddot{y},\ a_z = \dot{v}_z = \ddot{z},$$
$$\text{Magnitude} = a = \sqrt{a_x^2 + a_y^2 + a_z^2},\ \text{Direction: } \frac{\mathbf{a}}{a}.$$

Note. The equations describing the motion in *each coordinate direction* are identical, *in form*, to the equations that describe the motion of a point along a straight line. Consequently, you often can analyze the motion in *each coordinate direction* using the methods you applied to straight-line motion.

Motion of a Projectile

The free-flight motion of a projectile is often studied in terms of its Cartesian (rectangular) components, since the projectile's acceleration *always* acts in the vertical direction.

Angular Motion

- **Angular Motion of a Line**
  - **Angular Position.** We can specify the angular position of a line L in a particular plane relative to a reference line L_0 in the plane by an angle $\theta(rads)$.

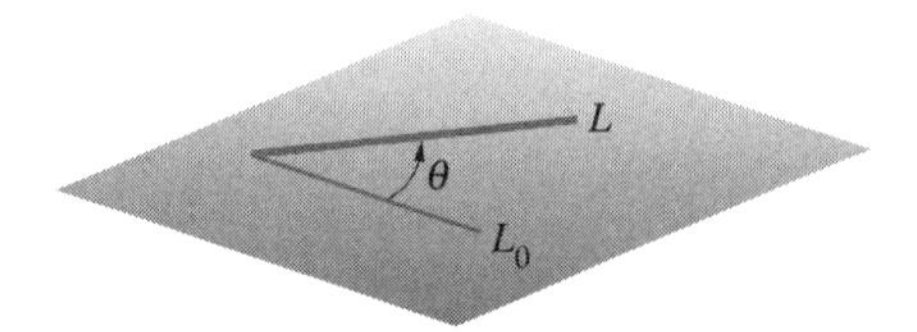

A line L and a reference line L_0 in a plane.

  - **Angular Velocity**. The *angular velocity* of L relative to L_0 is defined by

$$\omega = \frac{d\theta}{dt}\ (rad/s).$$

  - **Angular Acceleration.** The *angular acceleration* of L relative to L_0 is defined by

$$\alpha = \frac{d\omega}{dt} = \frac{d^2\theta}{dt^2}.\ (rad/s^2).$$

Normal and Tangential Components

In this method of describing curvilinear motion, we specify the position of a point by a coordinate measured along its path and express the velocity and acceleration in terms of their components tangential and normal to the path.

Important Points

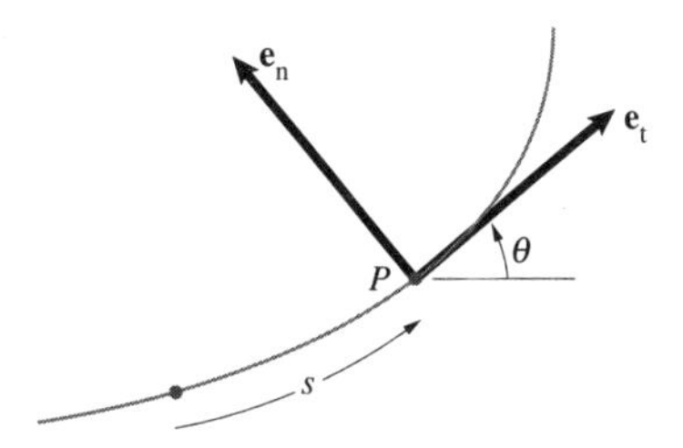

- Provided the *path* of the particle is *known*, we establish a set of n and t coordinates having a *fixed origin* which is coincident with the particle at the instant considered.
- The positive tangent axis always acts in the direction of motion and the positive normal axis is directed towards the path's center of curvature.
- The n and t axes are particularly advantageous for studying the velocity and acceleration of the particle, because the velocity $\mathbf{v}$ and the acceleration $\mathbf{a}$ are expressed by the following equations.
  - **Velocity.**

$$\mathbf{v} = v\mathbf{e}_t, \quad v = \frac{ds}{dt}.$$

  - **Acceleration.**

$$\mathbf{a} = a_t\mathbf{e}_t + a_n\mathbf{e}_n$$

  where

$$a_t = \frac{dv}{dt} \text{ and } a_n = v\frac{d\theta}{dt} = \frac{v^2}{\rho}.$$

  The unit vector $\mathbf{e}_n$ points toward the concave side of the path. The term ρ is the instantaneous radius of curvature of the path.

Circular Motion

If a point P moves in a plane *circular* path of radius R we have:

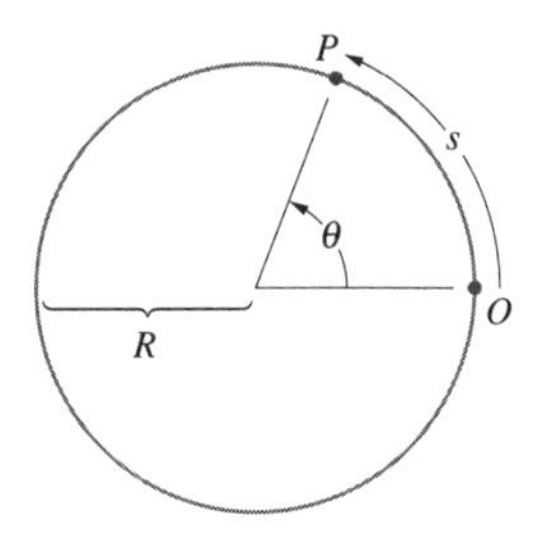

A point moving in a circular path.

- **Distance.**

$$s = R\theta.$$

- **Velocity.**

$$\mathbf{v} = R\omega\mathbf{e}_t. \tag{13.9}$$

- **Acceleration.**

$$\mathbf{a} = R\alpha\mathbf{e}_t + R\omega^2\mathbf{e}_n. \tag{13.10}$$

Polar and Cylindrical Coordinates

In some problems it is often convenient to express the path of motion in terms of cylindrical coordinates r, θ, z. If motion is restricted to the plane, polar coordinates r and θ are used.

1. Polar Coordinates

Coordinate System

- Polar coordinates are particularly suitable for solving problems for which data regarding the angular motion of the radial coordinate r is given to describe the particle's motion.
- To use polar coordinates, the origin is established at a fixed point, and the radial line r is directed to the particle.
- The transverse coordinate θ is measured counterclockwise from a fixed reference line to the radial line.

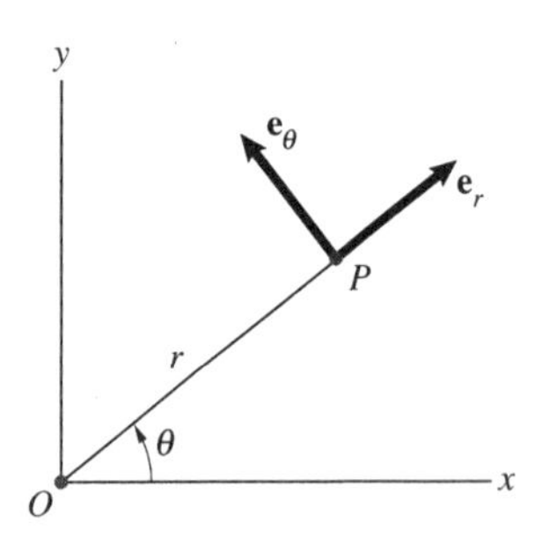

Position, Velocity, and Acceleration

- The *position* of the particle is defined by the position vector $\mathbf{r} = r\mathbf{e}_r$.
- The *velocity* and *acceleration* of the particle are given, respectively, by

$$\begin{aligned}\mathbf{v} &= v_r\mathbf{e}_r + v_\theta\mathbf{e}_\theta \\ &= \frac{dr}{dt}\mathbf{e}_r + r\omega\mathbf{e}_\theta,\end{aligned}$$

  and

$$\mathbf{a} = a_r\mathbf{e}_r + a_\theta\mathbf{e}_\theta,$$

where

$$a_r = \frac{d^2r}{dt^2} - r\left(\frac{d\theta}{dt}\right)^2 = \frac{d^2r}{dt^2} - r\omega^2 \quad \text{(radial component of acceleration)},$$

$$a_\theta = r\frac{d^2\theta}{dt^2} + 2\frac{dr}{dt}\frac{d\theta}{dt} = r\alpha + 2\frac{dr}{dt}\omega \quad \text{(transverse component of acceleration)}.$$

Circular Motion

If a point P moves in a plane *circular* path of radius R we have:

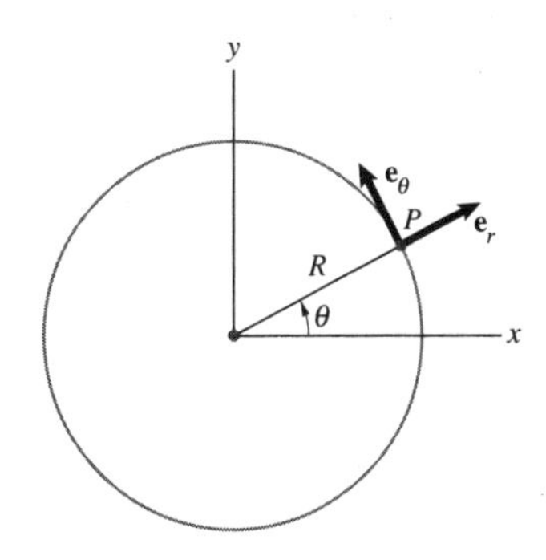

A point P moving in a circular path. (a) Polar coordinates. (b) Normal and tangential components.

- **Velocity.**

$$\mathbf{v} = R\omega\mathbf{e}_\theta. \tag{13.11}$$

- **Acceleration.**

$$\mathbf{a} = -R\omega^2\mathbf{e}_r + R\alpha\mathbf{e}_\theta. \tag{13.12}$$

2. Cylindrical Coordinates

Motion in *three-dimensions* requires a simple extension of the above formulas:

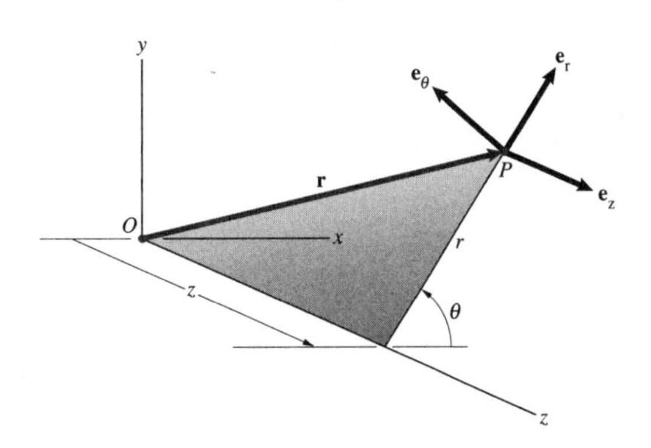

Cylindrical coordinates r, θ, and z of point P and the unit vectors e_r, e_θ, and e_z.

$$\text{Position: } \mathbf{r} = r\mathbf{e}_r + z\mathbf{e}_z,$$

$$\text{Velocity: } \mathbf{v} = \frac{dr}{dt}\mathbf{e}_r + r\omega\mathbf{e}_\theta + \frac{dz}{dt}\mathbf{e}_z,$$

$$\text{Acceleration: } a = (\frac{d^2r}{dt^2} - r\omega^2)\mathbf{e}_r + (r\alpha + 2\frac{dr}{dt}\omega)\mathbf{e}_\theta + \frac{d^2z}{dt^2}\mathbf{e}_z.$$

Helpful Tips and Suggestions

- Choose the coordinate system which is most convenient for the particular motion under consideration.
- Remember that the Eqs. (13.9)–(13.12) given above for circular motion are to be used *only* when the motion is *circular*.
- Notice that Eqs. (13.11) and (13.12) can be obtained from Eqs. (13.9) and (13.10) by noting that for circular motion

$$\mathbf{e}_\theta = \mathbf{e}_t \text{ and } \mathbf{e}_n = -\mathbf{e}_r.$$

- The most effective way to learn the principles of dynamics is to *solve problems* in a logical and orderly manner (*practice is the key!*).
- *Remember* that in solving problems from engineering mechanics you are solving real, practical problems and producing real data with physical significance. Thus, you are responsible for making sure your results are *correct, consistent* and *well presented.* Get into the habit of doing this *now* so that it will become second nature by the time you graduate. In the world of professional engineering, you have a *responsibility* to your profession and to the many people that will use the product you will help to design, manufacture, or implement.

Review Questions

1. Can the kinematics of a particle can be regarded as the same as the kinematics of a point?
2. Is the velocity of a point always tangent to its path ?
3. Is the acceleration of a point always tangent to its path?
4. If a ball is travelling in a horizontal circle at constant speed, does the center of the ball have zero acceleration?
5. Is the velocity of a point independent of the reference frame chosen to express the position of the point?
6. When does the acceleration of a particle always have a zero normal component?
7. If a particle in rectilinear motion has zero speed at some instant in time, is the acceleration necessarily zero at the same instant?
8. Can a particle have $\ddot{r} = 0$ but still have a nonvanishing radial component of acceleration?
9. If the magnitude of the velocity of a point moving in a curved path is constant, is the acceleration of the point zero?
10. What's the difference between *angular velocity* and *velocity of a point?*

14

Force, Mass, and Acceleration

Main Goals of this Chapter:

- To use Newton's second law to determine the acceleration of an object given information about the forces acting on the object.
- To use Newton's second law to obtain information about the forces acting an object given the object's acceleration.

14.1 Newton's Second Law

- *Newton's second law of motion* states that the *unbalanced force* on an object causes it to accelerate. If the mass of the particle is m and its velocity is $\mathbf{v}$, the second law can be written as:

$$\mathbf{f} = \frac{d}{dt}(m\mathbf{v}) \,. \tag{14.1}$$

  Further, if the mass m is constant, Eq. (14.1) becomes

$$\mathbf{f} = m\mathbf{a}. \tag{14.2}$$

  Equation (14.2) is referred to as the *equation of motion* and is one of the most important formulations in mechanics. Its validity is based *solely* on experimental evidence.
- Clearly, with the second law, a particle's motion can be determined when the total force acting on it is known, or the total force can be determined when the motion is known.

14.2 Equation of Motion for the Center of Mass

- Newton's second law is postulated for a particle, or small element of matter, but an equation of precisely the same form describes the motion of the center of mass of an arbitrary object. In fact

  The sum of the external forces acting on an object equals the product of the total mass and the acceleration of the center of mass

In other words, for the center of mass of an arbitrary object, the *equation of motion* (14.2) may be written as

$$\sum \mathbf{F} = m\mathbf{a}. \tag{14.3}$$

The magnitude and direction of each force acting on the object (left-hand side of Equation (14.3)) are identified using a *free-body diagram.*

14.3 Inertial Reference Frames

- Newton's second law *cannot* be expressed in terms of just *any* reference frame. Basically, the second law *requires* that the reference frame must not rotate and must either be fixed or translate in a given direction with *zero acceleration.* In many cases, however, the second law can still produce sufficiently accurate answers when applied with respect to a 'non-ideal' reference frame.
- We say that a reference frame in which Eq. (14.3) can be applied is said to be *Newtonian*, or *inertial.* In other words

  A reference frame is said to be inertial if it is one in which

  the second law (Eq. (14.2) or (14.3)) can be applied in this form.

- A reference frame translating at constant velocity relative to an inertial frame is also inertial.

14.4 Applications

To apply Newton's second law in a particular situation, a coordinate system must be chosen. Different types of coordinate systems can be used to analyze the different motions of objects and the forces acting on them.

Cartesian Coordinates and Straight-Line Motion

- When an object is moving relative to an inertial xyz frame of reference, the (vector) equation of motion (14.3) is equivalent to the following three *scalar* equations:

$$\begin{aligned} \sum F_x &= ma_x, \\ \sum F_y &= ma_y, \\ \sum F_z &= ma_z. \end{aligned} \tag{14.4}$$

  Only the first two of these equations are used to specify the motion of an object *constrained to move only in the* $x-y$ *plane.*

Normal and Tangential Components

- When an object moves over a *known planar curved path*, the equation of motion for the object may be written in the normal and tangential directions giving the following two *scalar* equations of motion:

$$\sum F_t = ma_t = m\frac{dv}{dt},$$

$$\sum F_n = ma_n = m\frac{v^2}{\rho}.$$

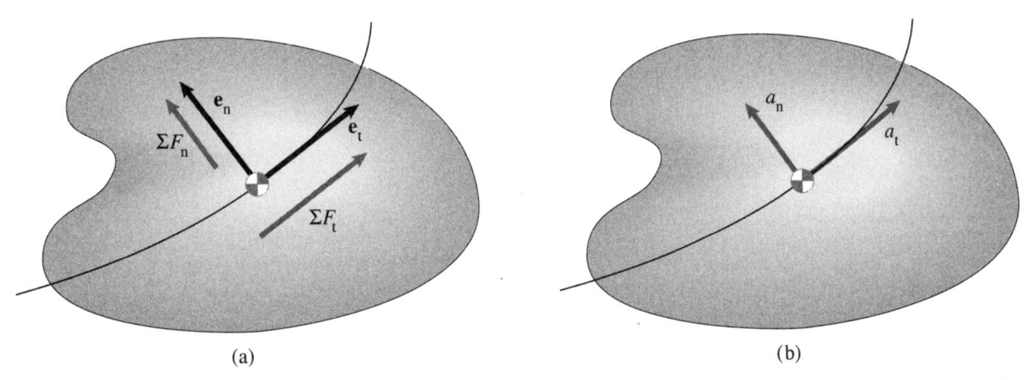

(a) Normal and tangential components of the sum of the forces on an object. (b) Normal and tangential components of the acceleration of the center of mass of the object.

- When an object moves in a *circular* path, normal and tangential components are usually the simplest choice for analyzing the motion of the object.

Polar and Cylindrical Coordinates

- When an object moves over a *known planar curved path*, the equation of motion for the object may be written in terms of polar coordinates giving the following two *scalar* equations of motion:

$$\sum F_r = ma_r = m\left(\frac{d^2r}{dt^2} - r\omega^2\right),$$

$$\sum F_\theta = ma_\theta = m\left(r\alpha + 2\frac{dr}{dt}\omega\right).$$

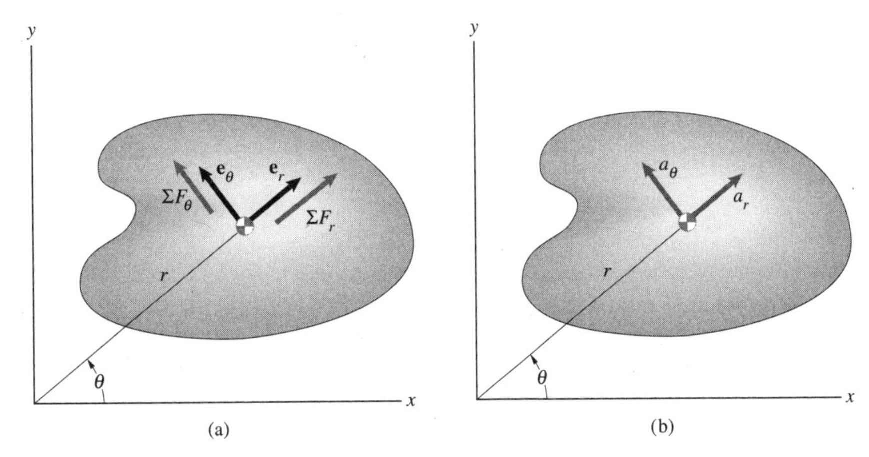

Radial and transverse components of (a) the sum of the forces and (b) the acceleration of the center of mass.

- The equation of motion for an object travelling over a *known three-dimensional curved path* may be written in terms of cylindrical coordinates r, θ, z giving the following three *scalar* equations of motion:

$$\sum F_r = ma_r = m\left(\frac{d^2r}{dt^2} - r\omega^2\right),$$

$$\sum F_\theta = ma_\theta = m\left(r\alpha + 2\frac{dr}{dt}\omega\right),$$

$$\sum F_z = ma_z = m\frac{dv_z}{dt} = m\frac{d^2z}{dt^2}.$$

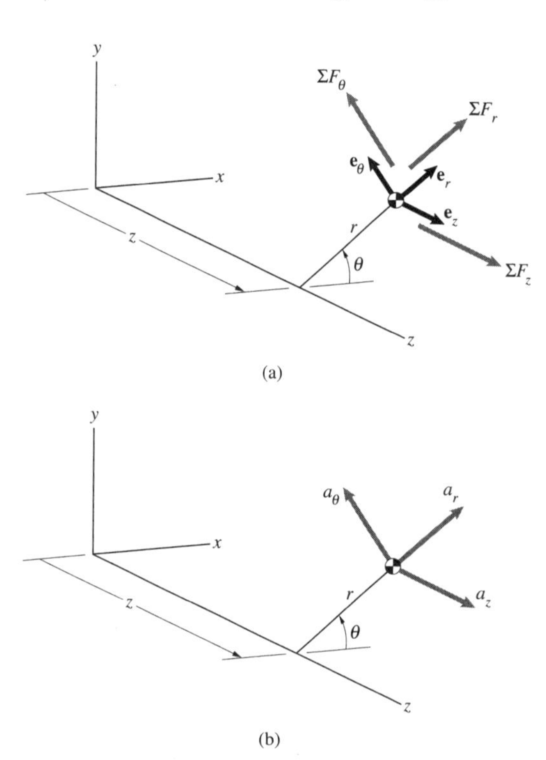

(a) Components of the sum of the forces on an object in cyclindrical coordinates. (b) Components of the acceleration of the center of mass.

Note:

- If the motion of an object is *confined to a fixed plane*, the component of the total force *normal to the plane* equals zero.
- In *straight-line motion*, the components of the total force *perpendicular to the line* equal zero and the component of the total force *tangent to the line* equals the product of the mass and the acceleration of the object *along the line*.

14.5 Orbital Mechanics

We can use Newton's second law expressed in polar coordinates to determine the orbit of an earth satellite or planet.

Determination of the Orbit

Suppose that at $t = 0$, a satellite has an initial velocity v_0 at a distance r_0 from the center of the earth. We assume that the initial velocity is perpendicular to the line from the center of the earth to the satellite. The satellite's position during its subsequent motion is specified by its polar coordinates (r, θ), where θ is measured from the satellite's position at $t = 0$.

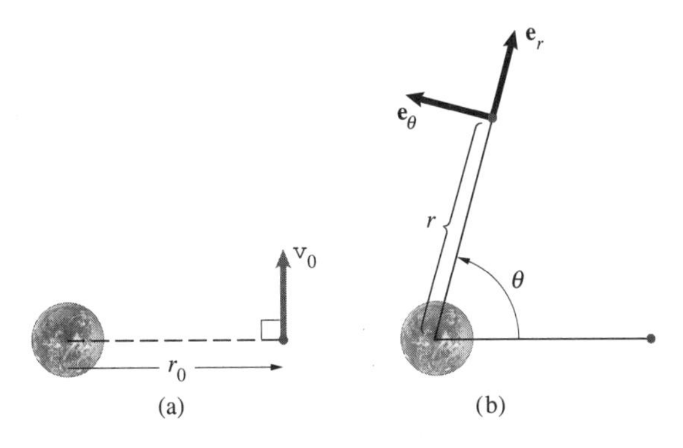

(a) Initial position and velocity of an earth satellite. (b) Specifying the subsequent path in terms of polar coordinates.

- The following linear differential equation defines the path $r = f(\theta)$ over which the satellite travels:

$$\frac{d^2u}{d\theta^2} + u = \frac{gR_E^2}{r_0^2 v_0^2}, \tag{14.5}$$

  where $u = \dfrac{1}{r}$ and R_E is the earth's radius.
- The initial conditions

$$\text{When } \theta = 0, u = \frac{1}{r_0} \quad \text{and} \quad v_r = \frac{dr}{dt} = 0 \text{ (so that } \frac{du}{d\theta} = 0)$$

  yield the following solution

$$\frac{r}{r_0} = \frac{1+\varepsilon}{1+\varepsilon\cos\theta}, \tag{14.6}$$

  where

$$\varepsilon = \frac{r_0 v_0^2}{gR_E^2} - 1. \tag{14.7}$$

- We note that Eq. (14.6) is the equation of a conic section with eccentricity ε in polar coordinates.
- The type of (conical) path taken by the satellite is determined from the value of the eccentricity ε of the conic section described by Eq. (14.6):

$\varepsilon = 0$	Orbit is circular and $v_0 = \sqrt{\dfrac{gR_E^2}{r_0}}$.
$0 < \varepsilon < 1$	Orbit is elliptic and $r_{max} = r_0\left(\dfrac{1+\varepsilon}{1-\varepsilon}\right)$.
$\varepsilon = 1$	Orbit is parabolic and $v_0 = \sqrt{\dfrac{2gR_E^2}{r_0}}$.
$\varepsilon > 1$	Orbit is a hyperbola.

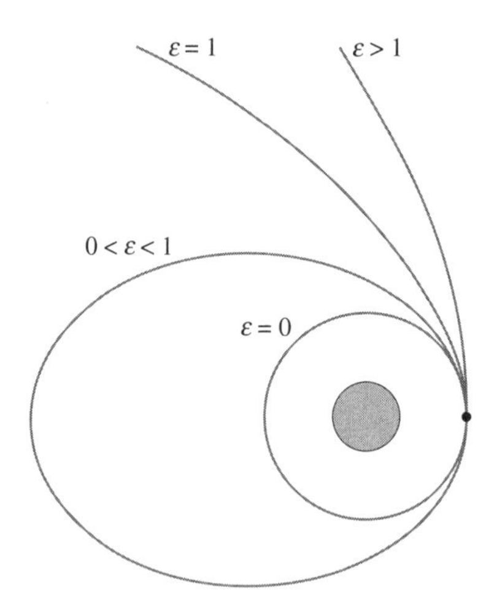

Orbits for different of eccentricity.

- Consequently, the initial launch velocity required for the satellite to follow a parabolic path ("on the border of never returning to its initial starting point") is called the *escape velocity*. The corresponding speed is given by

$$v_0 = \sqrt{\frac{2gR_E^2}{r_0}}.$$

- The speed v_0 required to launch a satellite into a circular orbit is given by

$$v_0 = \sqrt{\frac{gR_E^2}{r_0}}$$

  Speeds at launch less than v_0 will cause the satellite to reenter the earth's atmosphere and either burn up or crash.

- All trajectories attained by planets and most satellites are *elliptical*. For a satellite's orbit about the earth, the *minimum* distance r_0 from the orbit to the center of the earth is called the *perigee* of the orbit. The *maximum* distance is called the *apogee*.
- The fact that planets do indeed follow elliptic orbits about the sun was discovered by Johannes Kepler in the early seventeenth century—after 20 years of planetary observation! This led to *Kepler's laws for planetary motion*. Newton's *analytical verification* of Kepler's results was a triumph for Newtonian mechanics and was a confirmation of the inverse-square relation for gravitational force.

14.6 Numerical Solutions

Most problems that must be dealt with in engineering *cannot* be solved analytically with nice *closed-form solutions*. The functions describing the forces, and therefore the acceleration, are often *too complicated* to integrate and obtain closed-form solutions. In other situations, the forces are often not known in terms of functions, but instead are specified in terms of data, either as a continuous recording of force as a function of time (analog data) or as values of force measured at discrete times (digital data).

We can obtain approximate solutions to such problems by using *numerical integration*.

Helpful Tips and Suggestions

- Remember the importance of a free-body diagram in formulating the equation of motion—in any coordinate system.

Review Questions

1. True or False? The equation of motion is based solely on mathematical arguments.
2. Newton's second law states that the total external force on an object equals the product of the object's mass and the acceleration of which point?
3. What are the (scalar) equations of motion if an object is constrained to move only in the $x-y$ plane?
4. When is it preferable for the equation of motion of an object to be written in normal and tangential coordinates?
5. What method did Johannes Kepler use to discover that planets follow elliptic orbits around the sun?

15

Energy Methods

Main Goals of this Chapter:

- To develop the principle of work and energy and apply it to solve problems that involve force, velocity, and displacement.
- To introduce the concepts of work and power for an arbitrary object.
- To introduce the concept of a conservative force and apply the theorem of conservation of energy to solve kinetic problems.

15.1 Principle of Work and Energy

In mechanics a force **F** does work on an object only when the object undergoes a *displacement in the direction of the force*.

- The *work done* on an object as its center of mass moves from a position $\mathbf{r}_1$ to a position $\mathbf{r}_2$ is given by

$$U_{12} = \int_{\mathbf{r}_1}^{\mathbf{r}_2} \sum \mathbf{F} \cdot d\mathbf{r}, \tag{15.1}$$

  where $\sum \mathbf{F}$ is the sum of the forces acting on the object and $d\mathbf{r}$ is an infinitesimal displacement of the object's center of mass.
  Note. The term $\sum \mathbf{F} \cdot d\mathbf{r}$ is the *work*, expressed in terms of the total external force on the object and an infinitesimal displacement $d\mathbf{r}$ of its center of mass.
- The *principle of work and energy for an object* is given by

$$U_{12} = \frac{1}{2}mv_2^2 - \frac{1}{2}mv_1^2. \tag{15.2}$$

  The term $\frac{1}{2}mv^2$ is called the *kinetic energy* associated with the motion of the center of mass.

The work done on an object as it moves between two positions equals the change in its kinetic energy.

- The *principle of work and energy for a* **system of objects** states that:

> *The total work done by external forces on a system of objects as it moves between two positions equals the change in the total kinetic energy of the system if no net work is done by internal forces.*

> *The principle of work and energy is used to solve kinetic problems that involve velocity, force and displacement*

15.2 Work and Power

- **Evaluating the Work.** Let s be the position of an object's center of mass as it moves along its (curvilinear) path. Since, by definition, a force **F** does work on an object only when the object undergoes a *displacement in the direction of the force,* from Eq. (15.1), the work done in moving the object from a position $s = s_1$ to a position $s = s_2$ along its path is

$$\int_{s_1}^{s_2} \sum F_t ds$$

where $\sum F_t$ is the tangential component of the total external force on the object. In other words,

> *The work done is equal to the integral of the tangential component of the total force with respect to distance along its path.*

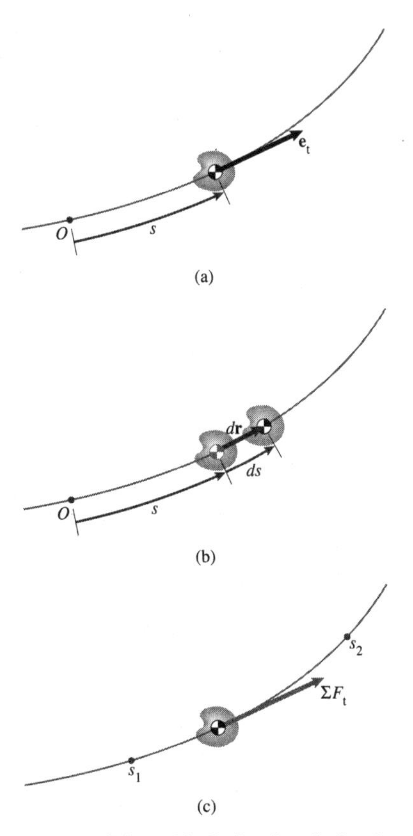

(a) The coordinate s and tangential unit vector. (b) An infinitesimal displacement $d\mathbf{r}$. (c) The work done from s_1 to s_2 is determined by the tangential component of the external forces.

Consequently,

— Components of force *perpendicular* to the path do no work.
— If $\sum F_t$ is constant between s_1 and s_2, the work is simply the *product of the total tangential force and the displacement.*
— If $\sum F_t$ is *opposite to the direction* of motion over some part of the path (the object is decelerating), the work is *negative.*

- **Power.** *Power* is the rate at which work is done. Hence, the power P generated by a machine or engine which performs an amount of work dU within a time interval dt, is given by

$$\begin{aligned} P &= \frac{dU}{dt} \\ &= \sum \mathbf{F} \cdot \frac{d\mathbf{r}}{dt} \\ &= \sum \mathbf{F} \cdot \mathbf{v} \end{aligned}$$

where $\mathbf{v}$ is the velocity of the point which is acted upon by the system of external forces $\sum \mathbf{F}$.
— Consequently power is a *scalar* with basic units *watt (W)* in the *SI* system and *horsepower (hp)* in the *FPS* system. Note that

$$\begin{aligned} 1W &= 1J/s = 1N \cdot m/s \\ 1hp &= 550 ft \cdot lb/s \\ 1hp &= 746W \end{aligned}$$

- The power equals the *rate of change of the object's kinetic energy.*
- The *average* (with respect to time) *power* transferred to or from an object during an interval of time from t_1 to t_2 is equal to the change in its kinetic energy, or the work done, divided by the interval of time:

$$P_{av} = \frac{\frac{1}{2}mv_2^2 - \frac{1}{2}mv_1^2}{t_2 - t_1} = \frac{U_{12}}{t_2 - t_1}.$$

15.3 Work Done by Particular Forces

In Section 15.2, you saw that if the tangential component of the total external force on an object is known as a *function of distance* along the object's path, the *principle of work and energy* can be used to relate a change in the position of the object to the change in its *velocity*.

For certain types of forces, however, not only can we determine the work *without knowing the tangential component* of the force as a function of distance along the path, but we don't even need to know the *path*! For example:

- **Weight.** In terms of a coordinate system with the positive *y axis* upward, the work done by an object's weight as the center of mass of the object moves from position 1 to position 2 is

$$U_{12} = -mg\,(y_2 - y_1).$$

— In other words, the work is the *product of the weight and the change in height of the center of mass of the object.*
— The work done is the *same* no matter what path the object follows from position 1 to position 2.

— The work is *negative* if the height *increases* and *positive* if it *decreases*.

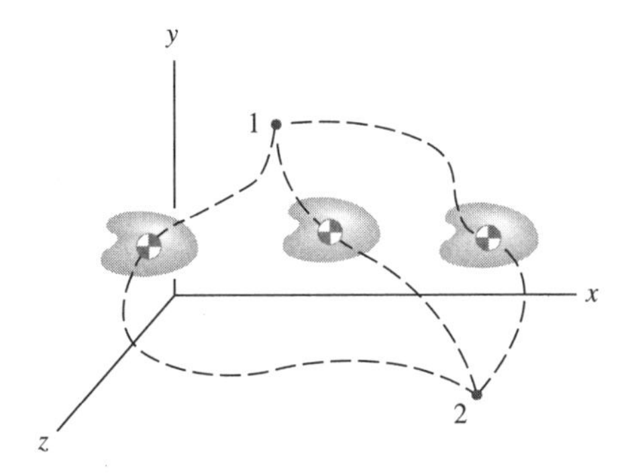

The work done by the weight is the same for any path.

— When the variation of an object's weight with distance r from the center of the earth is accounted for, the work done by the weight of the object is

$$U_{12} = mgR_E^2\left(\frac{1}{r_2} - \frac{1}{r_1}\right),$$

where R_E is the radius of the earth. Again the work is independent of path and all we need to evaluate it is the object's radial distance from the center of the earth at position 1 and position 2.

- **Springs.** The work done *on an object* by a spring attached to a fixed support is

$$U_{12} = -\frac{1}{2}k\left(S_2^2 - S_1^2\right),$$

where S_1 and S_2 are the values of the stretch at the initial and final positions of the object, respectively.
— Again, we don't need to know the object's path to determine the work done by the spring.

Potential Energy

15.4 Conservation of Energy

The work done on an object by some forces can be expressed as the change of a function of the object's position called the *potential energy*. When *all* of the forces that do work on a system have this property, we can state the principle of work and energy as a conservation law:

The sum of kinetic and potential energies is constant.

More precisely:

- **Potential Energy and Conservative Force.** For a given force $\mathbf{F}$ acting on an object, if a function V of the object's position exists such that

$$dV = -\mathbf{F} \cdot d\mathbf{r},$$

then $\mathbf{F}$ is said to be *conservative* and V is called the *potential energy* associated with the force $\mathbf{F}$.

- ♦ The work done by **F** from a position 1 to a position 2 is

$$U_{12} = -(V_2 - V_1).$$

- **Conservation of Energy.** If all the forces that do work on an object are *conservative*, the total energy—*the sum of the kinetic energy of the object and the potential energies of the forces acting on the object*—is constant or *conserved*:

$$\frac{1}{2}mv^2 + V = \text{constant}. \tag{15.3}$$

- ♦ If kinetic energy increases, V must decrease, and vice-versa, as if V represented a reservoir of "potential" kinetic energy (hence the term "potential energy").
- ♦ When Eq. (15.3) is true, the system is said to be *conservative* and we can use Eq. (15.3) *instead* of the principle of work and energy to relate a change in position of the system to the change in its kinetic energy. The two approaches are *equivalent*, but greater insight is gained by using conservation of energy.

The conservation of energy equation (15.3) is used to solve problems involving velocity,displacement and conservative force systems all of which form part of the energy equation (15.3)

15.5 Conservative Forces

- **Conservative Force.** When the work done by a force in moving an object from one point to another is *independent of the path* followed by the object, this force is called a *conservative force.* e.g.
  - — The work done by the *weight of an object* is *independent of the path* of the object i.e., the work done depends only on the object's *vertical displacement.*
  - — The work done by a spring force *acting on an object* is *independent of the path* of the object i.e., it depends only on the extension or compression of the spring.
  - — In contrast, we note that the *force of friction* exerted *on a moving object* by a fixed surface *depends on the path* of the object i.e., the longer the path, the greater the work. Consequently, frictional forces are *nonconservative*. The work is dissipated from the body in the form of heat.

Potential Energies of Particular Forces

- **Weight.** In terms of a Cartesian coordinate system with its y *axis* pointing upward, the potential energy of the weight of an object is

$$V = mgy,$$

where y is the height of the object's mass center, measured *positive upward* from an arbitrarily selected reference level or *datum*.
Note. When the variation of an object's weight with distance r from the center of the earth is accounted for, the potential energy of the weight is

$$V = -\frac{mgR_E^2}{r},$$

where R_E is the radius of the earth.

- **Springs.** The elastic potential energy due to a spring's configuration (stretched or compressed a distance S from its unstretched position) is

$$V = \frac{1}{2}kS^2.$$

Note that V is *always positive* since, in the deformed position, the force of the spring has the capacity for always doing positive work on the object when the spring is returned to its unstretched position.

Solving Problems Using the Conservation of Energy (15.3)

The conservation of energy equation is used to solve problems involving *velocity, displacement and conservative force systems*. It is generally *easier to apply* than the principle of work and energy because the energy equation just requires specifying the particle's kinetic and potential energies at only *two points* along the path, rather than determining the work done when the particle moves through a displacement. The procedure is as follows:

1. *Determine whether the system is conservative.* Draw a free-body diagram to identify the forces that do work, and confirm that they are conservative.
2. *Determine the potential energy*. Evaluate the potential energies of the forces in terms of the position of the system.
3. *Apply conservation of energy*. Equate the sum of the kinetic and potential energies of the system at two positions to obtain an expression for the change in the kinetic energy.

15.6 Relationships between Force and Potential Energy

- A force $\mathbf{F}$ is related to its associated potential energy V by

$$\mathbf{F} = -\left(\frac{\partial V}{\partial x}\mathbf{i} + \frac{\partial V}{\partial y}\mathbf{j} + \frac{\partial V}{\partial z}\mathbf{k}\right) = -\nabla V.$$

- A force $\mathbf{F}$ is conservative if and only if

$$curl\mathbf{F} = \nabla \times \mathbf{F} = \begin{vmatrix} \mathbf{i} & \mathbf{j} & \mathbf{k} \\ \frac{\partial}{\partial x} & \frac{\partial}{\partial y} & \frac{\partial}{\partial z} \\ F_x & F_y & F_z \end{vmatrix} = \mathbf{0}.$$

Helpful Tips and Suggestions

- Only problems involving *conservative forces* (e.g., weights and springs) may be solved using *conservation of energy* (Eq. (15.3)). Friction or other drag-resistant forces, which depend on velocity or acceleration are nonconservative (a portion of the work done by such forces is transformed into thermal energy which dissipates into the surroundings and may not be recovered). When such forces enter into the problem, use the *principle of work and energy*.

Review Questions

1. How would you calculate the work done on an object by external forces $\sum \mathbf{F}$?
2. When is the principle of work and energy used to solve kinetic problems?

3. If an object is subjected only to forces that are perpendicular to its path, what do you know about its kinetic energy?
4. How is power defined and how is it calculated?
5. What is a conservative force? Give some examples of conservative forces.
6. Explain why the weight of an object is a conservative force.
7. Give an example of a nonconservative force and explain why the force is nonconservative.
8. What is potential energy ? Give some examples.
9. How can you prove that a force is conservative?
10. When is the conservation of energy equation used to solve problems in kinetics?

16

Momentum Methods

Main Goals of this Chapter:

- To develop the principle of linear impulse and momentum for an object.
- To study the conservation of linear momentum.
- To analyze the mechanics of impact.
- To introduce the concept of angular impulse and momentum.
- To solve problems involving continuous flows of mass.

16.1 Principle of Impulse and Momentum

- The *principle of impulse and momentum* is obtained from a time integration of the equation of motion and is stated as follows:

  The linear impulse applied to an object during an interval of time is equal to the change in the object's linear momentum.

$$\int_{t_1}^{t_2} \sum \mathbf{F} dt = m\mathbf{v}_2 - m\mathbf{v}_1, \tag{16.1}$$

  where $\mathbf{v}_1$ and $\mathbf{v}_2$ are the velocities of the center of mass of the object at the times t_1 and t_2. The term on the left-hand side of Eq. (16.1) is called the *linear impulse* and $m\mathbf{v}$ is called the *linear momentum of the object.*
- When the external forces acting on an object are known as *functions of time*, the principle of impulse and momentum may be applied to determine the *change in velocity of the object during an interval of time.*
- This result can also be expressed in terms of the average of the total force with respect to time:

$$(t_2 - t_1) \sum \mathbf{F}_{\mathrm{av}} = m\mathbf{v}_2 - m\mathbf{v}_1.$$

- The *principle of impulse and momentum* can also be expressed in terms of the tangential component of the total force and the linear momentum along the object's path:

$$\int_{t_1}^{t_2} \sum F_t \, dt = mv_2 - mv_1.$$

- The average of the tangential component of the total force with respect to time is related to the change in the linear momentum along the path by

$$(t_2 - t_1) \sum F_{t\,\text{av}} = mv_2 - mv_1.$$

Comparison Between Principles of Impulse and Momentum and Work and Energy

1. Both principles relate an integral of the external forces to the change in an object's velocity.
2. The principle of impulse and momentum requires that the external forces acting on the object are known as functions of time. The principle of work and energy requires that the external forces be known as functions of position.
3. Eq. (16.1) is a *vector* equation that determines the change in *both the magnitude and direction* of the velocity, whereas the principle of work and energy is a scalar equation which gives only the change in the *magnitude* of the velocity.
4. In the case of impulse and momentum, there is no class of forces equivalent to the conservative forces that make the principle of work and energy so easy to apply.

16.2 Conservation of Linear Momentum

- If objects A and B are not subjected to external forces other than the forces they exert on each other (or if the effects of other external forces are *negligible*), their total linear momentum is conserved:

$$m_A \mathbf{v}_A + m_B \mathbf{v}_B = \text{constant}. \tag{16.2}$$

- Also, the velocity of the combined center of mass of the objects A and B (that is, the center of mass of A and B regarded as a single object) is constant.
- Even when significant external forces act on A and B, if the external forces are negligible in a particular direction, Eq. (16.2) applies in that direction.
- Eq. (16.2) also applies to an arbitrary number of objects:

If the external forces acting on any collection of objects are negligible, the total linear momentum is conserved and the velocity of their center of mass is conserved.

16.3 Impacts

- If colliding objects A and B are not subjected to external forces, *their total linear momentum must be the same before and after the impact* (total linear momentum of the system composed of objects A and B is *conserved*).
- Even when the colliding objects A and B are subjected to external forces, the force of the impact is often so large, and its duration so brief, that the effect of the external forces on their motions during the impact is *negligible*.

- If objects A and B adhere and remain together after they collide, they are said to undergo a *perfectly plastic impact*. The velocity of their common center of mass before and after the impact is given by

$$\mathbf{v} = \frac{m_A \mathbf{v}_A + m_B \mathbf{v}_B}{m_A + m_B}. \tag{16.3}$$

A remarkable feature of this result is that we determine the velocity following the impact *without considering the physical nature of the impact.*

- If objects A and B do not adhere, linear momentum conservation alone does not provide enough equations to determine their velocities after impact.

Direct Central Impact

- Suppose that the centers of mass of objects A and B travel along the same straight line with velocities v_A and v_B before their impact and with velocities v'_A and v'_B along the same straight line after their impact. In this *direct central impact*, linear momentum is *conserved*, or

$$m_A v_A + m_B v_B = m_A v'_A + m_B v'_B,$$

and the velocities are related by the *coefficient of restitution*:

$$e = \frac{v'_B - v'_A}{v_A - v_B}.$$

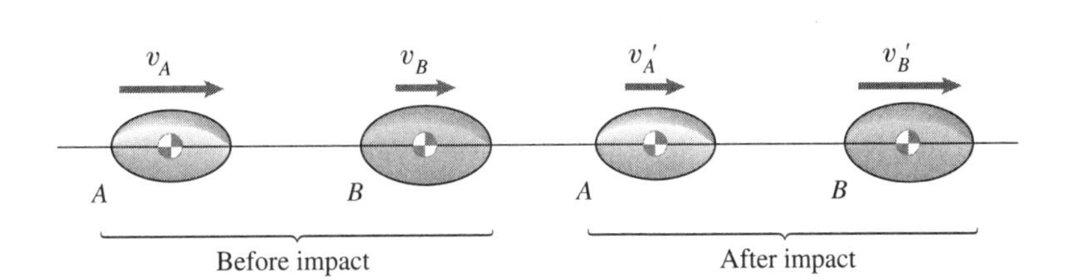

- **Perfectly Elastic Impact**: $e = 1$. Total kinetic energy is conserved
- **Perfectly Plastic Impact**: $e = 0$. Both objects *stick together* and move with a common velocity.

Oblique Central Impact

- Suppose that objects A and B approach with arbitrary velocities $\mathbf{v}_A$ and $\mathbf{v}_B$ and that the forces they exert on each other during their impact are parallel to the x *axis* and point toward their centers of mass. In this *oblique central impact*, no forces are exerted on A and B in the y and z directions, so the components of velocity in the y and z directions are unchanged by the impact:

$$(\mathbf{v}'_A)_y = (\mathbf{v}_A)_y, \quad (\mathbf{v}'_B)_y = (\mathbf{v}_B)_y,$$
$$(\mathbf{v}'_A)_z = (\mathbf{v}_A)_z, \quad (\mathbf{v}'_B)_z = (\mathbf{v}_B)_z.$$

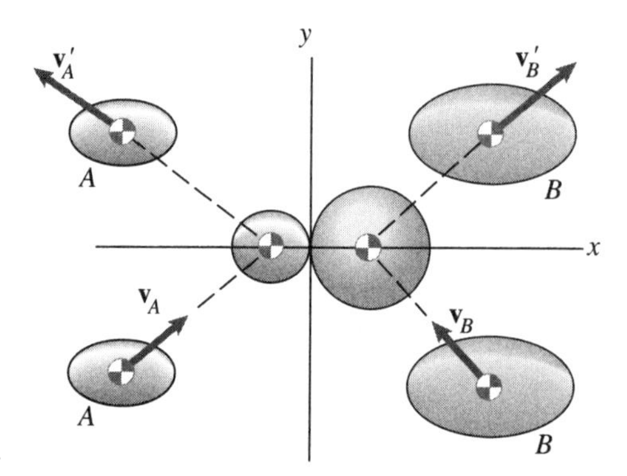

- In the *x direction*, linear momentum is conserved or

$$m_A (\mathbf{v}_A)_x + m_B (\mathbf{v}_B)_x = m_A \left(\mathbf{v}'_A\right)_x + m_B \left(\mathbf{v}'_B\right)_x$$

and the velocity components are related by the coefficient of restitution

$$e = \frac{\left(\mathbf{v}'_B\right)_x - \left(\mathbf{v}'_A\right)_x}{(\mathbf{v}_A)_x - (\mathbf{v}_B)_x}.$$

- **Note.** The *principle of work and energy* cannot be used to analyze impact problems since it is not possible to know how the *internal forces* of deformation and restitution vary or displace during the collision.

16.4 Angular Momentum

- **Principle of Angular Impulse and Momentum.**

For a fixed point O, the angular impulse applied to an object during an interval of time is equal to the change in the object's angular momentum about O.

That is,

$$\int_{t_1}^{t_2} \left(\mathbf{r} \times \sum \mathbf{F}\right) dt = (\mathbf{H}_O)_2 - (\mathbf{H}_O)_1, \tag{16.4}$$

where the angular momentum about O (or the moment of the linear momentum of the object about point O) is

$$\mathbf{H}_O = \mathbf{r} \times m\mathbf{v},$$

and the integral on the left-hand side of (16.4) is called the *angular impulse*.

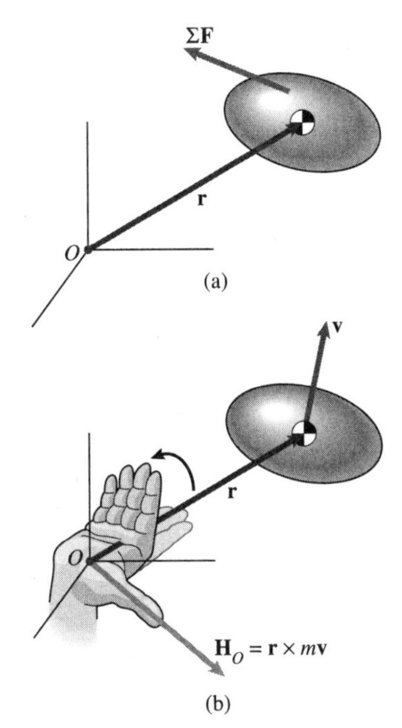

(a) The position vector and the total external force on an object. (b) The angular momentum vector and the right-hand rule for determining its direction.

- **Central-Force Motion.** If the total force acting on an object remains directed toward a point O that is fixed relative to an inertial reference frame, the object is said to be in *central-force motion* (e.g., orbits are examples of central-force motion). The fixed point is called the *center of the motion*. In central-force motion, the angular momentum of the object about the fixed point is conserved:

$$\mathbf{H}_O = \text{constant}.$$

♦ **Plane Central-Force Motion.** In plane central-force motion, the product of the radial distance from the center of motion and the transverse component of the velocity is constant:

$$r v_\theta = \text{constant}.$$

16.5 Mass Flows

Conservation of linear momentum can be used to determine the force exerted on an object as a result of emitting or absorbing a continuous flow of mass:

- A flow of mass *from* an object with velocity $\mathbf{v}_f$ *relative to the object* exerts a force

$$\mathbf{F}_f = -\frac{dm_f}{dt}\mathbf{v}_f$$

on the object, where $\frac{dm_f}{dt}$ is the *mass flow rate*.

- The direction of the force is *opposite* to the direction of the relative velocity.
- A flow of mass *to* an object exerts a force in the *same* direction as the relative velocity.

Helpful Tips and Suggestions

- Unlike energy, momentum *is a vector* and so has both magnitude and direction.
- Be familiar with the main differences between the principles of linear impulse and momenum and work and energy—so that you know *which* one to use and *when*.

Review Questions

1. What is the linear momentum of an object?
2. When can the principle of *linear impulse and momentum* be applied and why ?
3. How does the principle of impulse and momentum differ from the principle of work and energy?
4. When does the principle of linear impulse and momentum become *conservation of linear momentum*?
5. If the only forces exerted on two objects are the forces they exert on each other, what can be inferred about the motion of their combined center of mass?
6. If two objects collide, what can you infer about the motion of their common center of mass?
7. Define the *angular momentum* of an object about a point O.
8. What is meant by conservation of angular momentum ?
9. True or false? If an object's angular momentum is conserved, the object's linear momentum must also be conserved.
10. In central-force motion, what can be inferred about the angular momentum?

17

Planar Kinematics of Rigid Bodies

Main Goals of this Chapter:

- To classify the various types of rigid-body motion.
- To analyze the motion of objects rotating about a fixed axis.
- To provide a relative motion analysis of velocity and acceleration.
- To show how to find the instantaneous center of zero velocity and determine the velocity of a point on a body using this method.
- To examine a class of problems dealing with sliding contacts.
- To provide a relative motion analysis of velocity and acceleration using a moving frame of reference.

17.1 Rigid Bodies and Types of Motion

- **Rigid Body.** A rigid body is an idealized model of an object that does not deform or change shape. In other words,

The distance between every pair of points of a rigid body remains constant.

- **Types of Motion**
  1. **Translation.** If a rigid body in motion relative to a given reference frame *does not rotate*, it is said to be in translation:
     - **(i) Rectilinear Translation**. All points on the body follow parallel straight-line paths.
     - **(ii) Curvilinear Translation.** All points on the body follow curved paths that are the same shape and are equidistant from one another.

Every point of a rigid body in translation has the same velocity and acceleration so that its motion can be described completely if we describe only the motion of a single point.

2. **Rotation about a Fixed Axis.** All of the points of the body, except those which lie on the axis of rotation, move along circular paths.
3. **Planar Motion**. If the points of a rigid body intersected by a fixed plane remain in that plane, the rigid body is said to undergo two-dimensional or *planar motion*. Rotation of a rigid body about a fixed axis is a special case of planar motion.

17.2 Rotation about a Fixed Axis

Consider a rigid body rotating about a fixed axis. The angle θ between the reference line and the body-fixed line describes the position or orientation of the rigid body about the fixed axis.

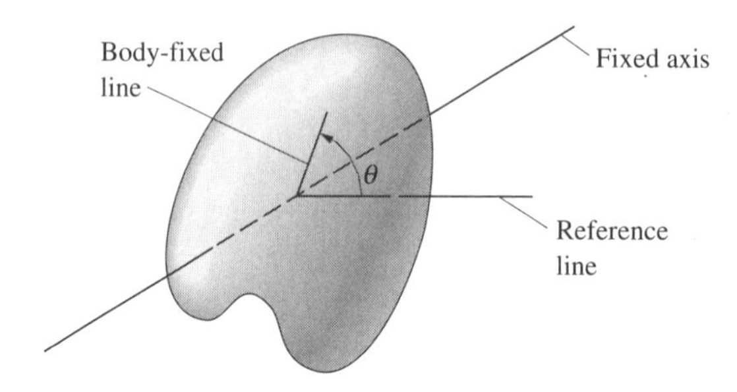

Specifying the orientation of an object rotating about a fixed axis.

- The rigid body's angular velocity (or rate of rotation) and is angular acceleration are described by

$$\omega = \frac{d\theta}{dt}, \quad \alpha = \frac{d\omega}{dt} = \frac{d^2\theta}{dt^2}.$$

- Since each point of the object moves in a circular path about the axis of rotation, the magnitude of the velocity of a point at a distance r form the fixed axis is

$$v = r\omega.$$

The components of acceleration are

$$a_t = r\alpha, \quad a_n = \frac{v^2}{r} = r\omega^2.$$

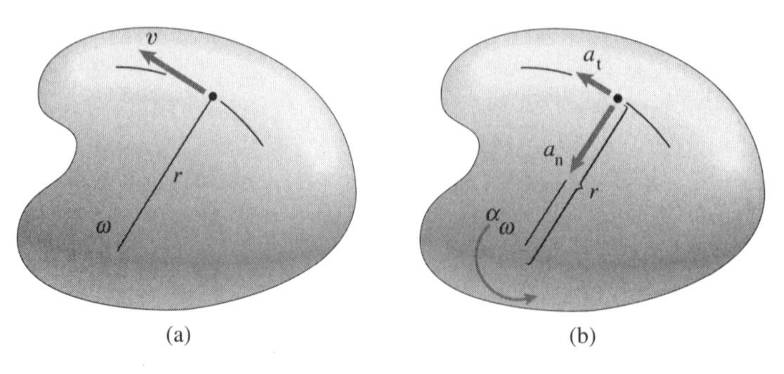

(a) Velocity and (b) acceleration of a point of a rigid body rotating about a fixed axis.

17.3 General Motions: Velocities

The Angular Velocity Vector

- The angular velocity vector ω of a rigid body is parallel to the *instantaneous axis of rotation of the body*, and its magnitude $|\omega|$ is the body's *rate of rotation.*
- If the thumb of the right hand points in the direction of ω, the fingers curl around ω in the direction of the rotation.

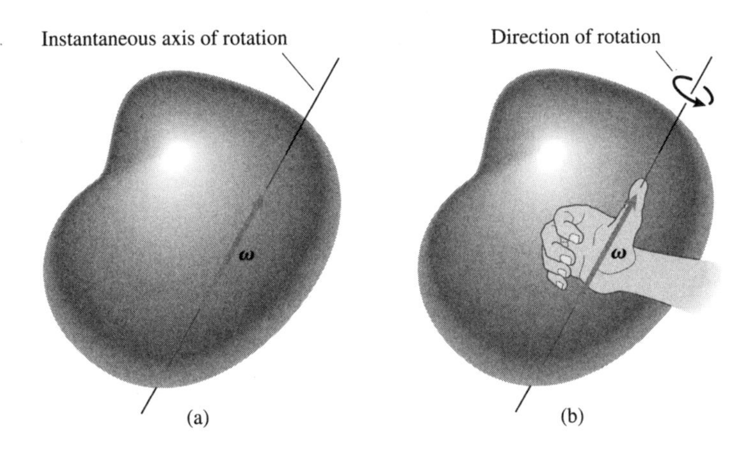

(a) An angular velocity vector. (b) Right-hand rule for the direction of the vector.

Relative Velocities

- Suppose O is the origin of a given reference frame. Suppose further that the velocity $\mathbf{v}_B$ of any point B on the rigid body and the body's angular velocity ω are known relative to the given reference frame. Then, the relative velocity equation

$$\begin{aligned}\mathbf{v}_A &= \mathbf{v}_B + \mathbf{v}_{A/B} \\ &= \mathbf{v}_B + \omega \times \mathbf{r}_{A/B} \qquad (17.1)\end{aligned}$$

can be used to determine the velocity of any other point A on the rigid body. Here, $\mathbf{r}_{A/B}$ is the position vector of point A *relative to point* B (as if point B was the origin of the coordinate system)

Instantaneous Centers

- A point of a rigid body whose velocity is zero at a given instant is called an *instantaneous center*. "Instantaneous" means that the point may have zero velocity *only* at the instant under consideration.
- When we know the location of an instantaneous center of a rigid body in two-dimensional motion and we know the body's angular velocity, the velocities of other points are easy to determine. To see this, let C be the location of the instantaneous center. Then $\mathbf{v}_C = \mathbf{0}$, and

$$\begin{aligned}\mathbf{v}_A &= \mathbf{v}_C + \omega \times \mathbf{r}_{A/C} \\ &= \omega \times \mathbf{r}_{A/C}\end{aligned}$$

Hence, point A moves momentarily about the instantaneous center in a *circular path.*

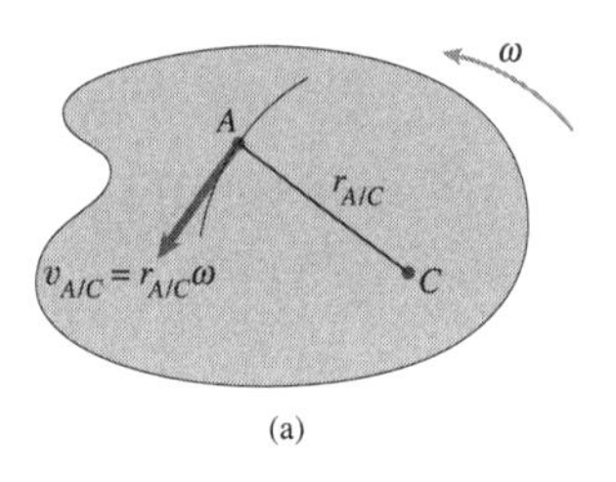

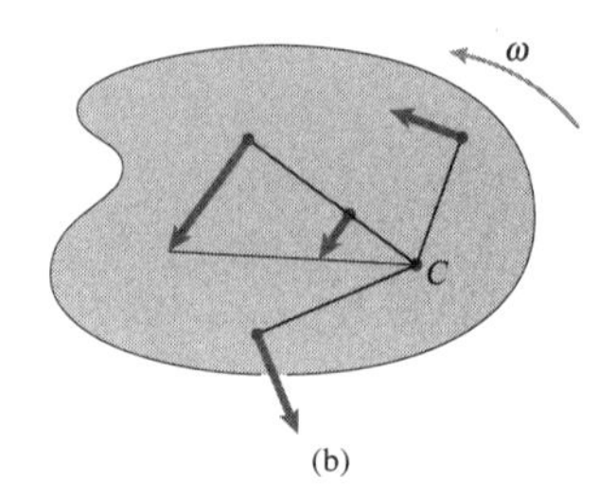

(a) An instantaneous center C and a different point A. (b) Every point is rotating about the instantaneous center.

Location of the Instantaneous Center

Given two nonparallel velocities $\mathbf{v}_A$ **and** $\mathbf{v}_B$. Construct at points A and B line segments that are perpendicular to $\mathbf{v}_A$ and $\mathbf{v}_B$. Extending these perpendiculars to their *point of intersection* as shown locates the instantaneous center at the instant considered.

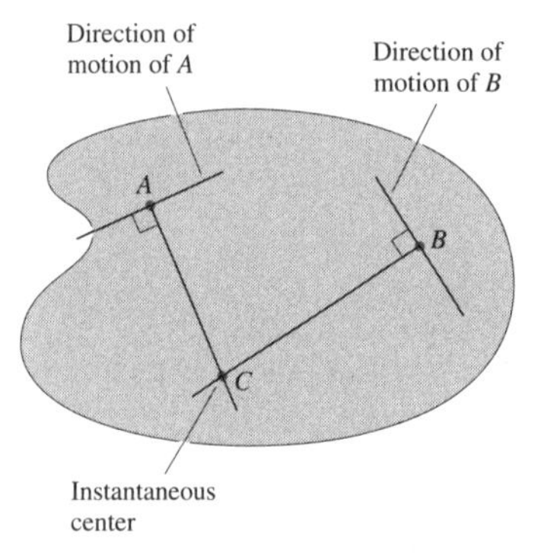

Locating the instantaneous center in planar motion.

Note. The instantaneous center may not be a point of the rigid body. This simply means that at the instant in question, the rigid body is rotating about an external point. It is helpful to imagine *extending* the rigid body so that it includes the instantaneous center.

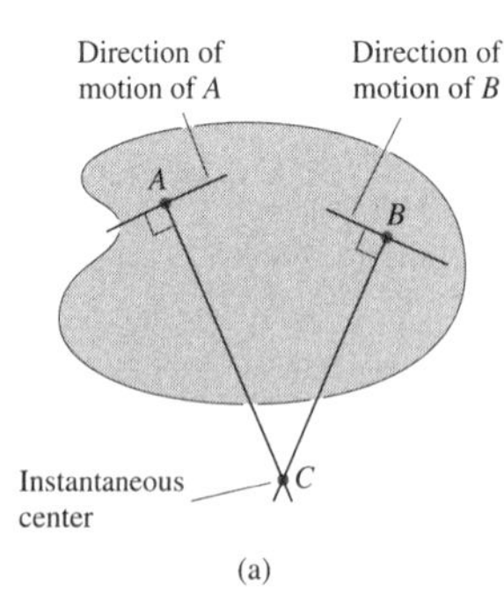

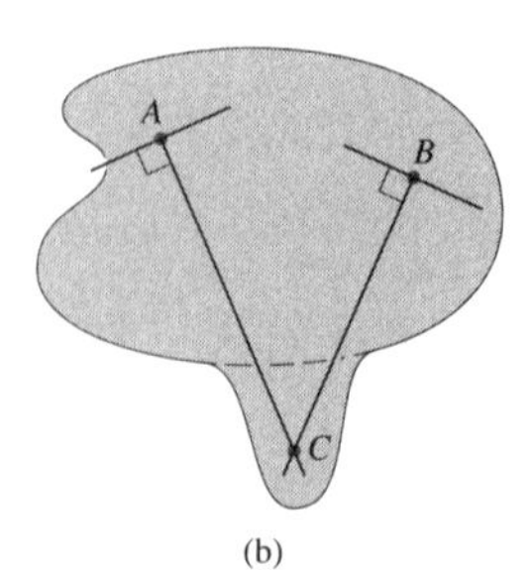

(a) An instantaneous center external to the rigid body. (b) A hypothetical extended body. Point C would be stationary.

Important Notes

1. The point chosen as the instantaneous center for the body can be used *only for an instant of time* since the body changes its position from one instant to the next.
2. The instantaneous center *does not, in general, have zero acceleration* and so *should not* be used for finding accelerations of points in a body.

3. If the location of the instantaneous center is "at infinity" (velocities of points A and B are parallel), the rigid body is in pure translation, with an angular velocity of zero.

17.4 General Motions: Acceleration

- The angular acceleration vector $\alpha = \dfrac{d\omega}{dt}$ is the *rate of change of the angular velocity vector* of the body.
- An equation relating the accelerations of two points on a rigid body is given by

$$\mathbf{a}_A = \mathbf{a}_B + \alpha \times \mathbf{r}_{A/B} + \omega \times \left(\omega \times \mathbf{r}_{A/B}\right).$$

An equation relating the accelerations of two points on a rigid body *subjected to planar motion* is given by:

$$\mathbf{a}_A = \mathbf{a}_B + \alpha \times \mathbf{r}_{A/B} - \omega^2 \mathbf{r}_{A/B} \tag{17.2}$$

where, relative to a given reference frame,

$\mathbf{a}_A =$ acceleration of point A

$\mathbf{a}_B =$ acceleration of point B

$\alpha =$ angular acceleration of the body

$\omega =$ angular velocity of the body

$\mathbf{r}_{A/B} =$ relative-position vector drawn from B to A

- **Important Point:** Before applying Eq. (17.2), it will be necessary to determine the angular velocity ω of the body by using a velocity analysis and Eq. (17.1).

17.5 Sliding Contacts

In the case of sliding contacts, a slightly different method of solution is required in which cannot assume that the point A in Eqs (17.1) and (17.2) is a point of the rigid body.

- Consider then a point B of a rigid body, a *body-fixed* reference frame (one that moves with the rigid body) with origin at B and an arbitrary point A.

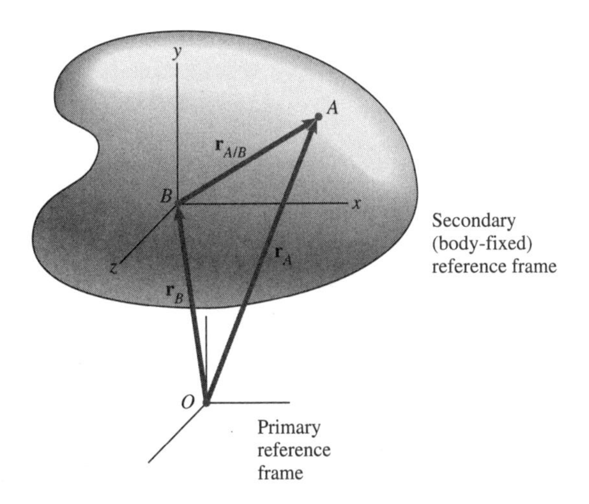

A point B of a rigid body, a body-fixed secondary reference frame, and an arbitrary point A.

- The velocities $\mathbf{v}_A$ and $\mathbf{v}_B$ of the points relative to the *primary* reference frame are related by

$$\mathbf{v}_A = \mathbf{v}_B + (\mathbf{v}_A)_{\text{rel}} + \omega \times \mathbf{r}_{A/B},$$

  where $(\mathbf{v}_A)_{\text{rel}}$ is the velocity of A relative to the body-fixed reference frame.
  Note. If *A is a point of the rigid body*, $(\mathbf{v}_A)_{\text{rel}}$ is zero.
- The accelerations $\mathbf{a}_A$ and $\mathbf{a}_B$ of the points relative to the primary reference frame are related by

$$\mathbf{a}_A = \mathbf{a}_B + (\mathbf{a}_A)_{\text{rel}} + 2\omega \times (\mathbf{v}_A)_{\text{rel}} + \alpha \times \mathbf{r}_{A/B} + \omega \times \left(\omega \times \mathbf{r}_{A/B}\right) \tag{17.3}$$

  where $(\mathbf{a}_A)_{\text{rel}}$ is the acceleration of A relative to the *body-fixed* reference frame.
  - In planar motion, the term $\omega \times \left(\omega \times \mathbf{r}_{A/B}\right)$ can be written in the simpler form $-\omega^2\mathbf{r}_{A/B}$.
    Note. If *A is a point of the rigid body*, $(\mathbf{a}_A)_{\text{rel}}$ is zero.

17.6 Moving Reference Frames

In many situations, it is convenient to describe the motion of a point by using a *secondary reference frame* that moves *relative* to some *primary reference frame.*

- Consider a point A and a reference frame, with origin at B, that rotates with angular velocity ω and angular acceleration α *relative to a primary reference frame.*

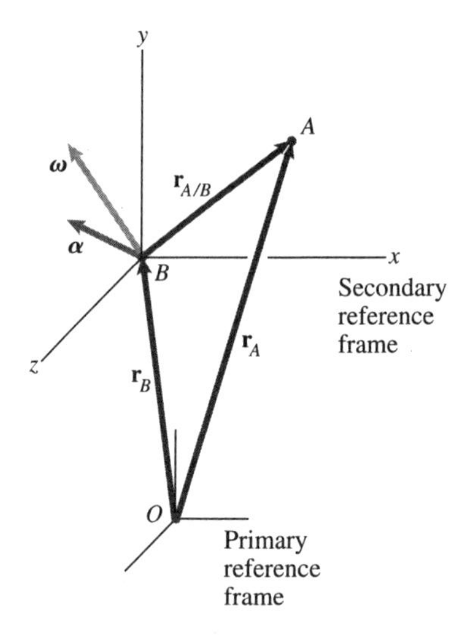

A secondary reference frame with origin B and an arbitrary point A.

- The velocities of A and B *relative to the primary reference frame* are related by

$$\mathbf{v}_A = \mathbf{v}_B + (\mathbf{v}_A)_{\text{rel}} + \omega \times \mathbf{r}_{A/B}$$

  where $(\mathbf{v}_A)_{\text{rel}}$ is the velocity of A relative to the secondary reference frame.
- The accelerations of A and B *relative to the primary reference frame* are related by

$$\mathbf{a}_A = \mathbf{a}_B + (\mathbf{a}_A)_{\text{rel}} + 2\omega \times (\mathbf{v}_A)_{\text{rel}} + \alpha \times \mathbf{r}_{A/B} + \omega \times \left(\omega \times \mathbf{r}_{A/B}\right) \tag{17.4}$$

  where $(\mathbf{a}_A)_{\text{rel}}$ is the acceleration of A relative to the secondary reference frame.

Inertial Reference Frames

- We say that a reference frame is *inertial* if it can be used to apply Newton's second law in the form $\sum \mathbf{F} = m\mathbf{a}$.
- In many cases we treat reference frames as inertial *even though* they both accelerate and rotate. This is usually explained by the fact that the "errors" introduced in doing so are *negligible* in specific cases. For example:
  - **Earth-Centered, Nonrotating Reference Frame.** A nonrotating reference frame with its origin at the center of the earth can be assumed to be inertial for the purpose of describing motions of *objects near the earth.*
  - **Earth-Fixed Reference Frame.** A local, earth-fixed reference frame can also be assumed to be inertial for the purpose of describing motions of *objects on or near the earth.*
- **Coriolis Effects.** The *Coriolis force* $2m\omega \times (\mathbf{v}_A)_{\text{rel}}$ explains a number of physical phenomena that exhibit different behaviors in the northern and southern hemispheres.
- **Arbitrary Reference Frame.** Suppose that the primary reference frame with its origin at O is inertial and the secondary reference frame with its origin at B undergoes an arbitrary motion with angular velocity ω and angular acceleration α.

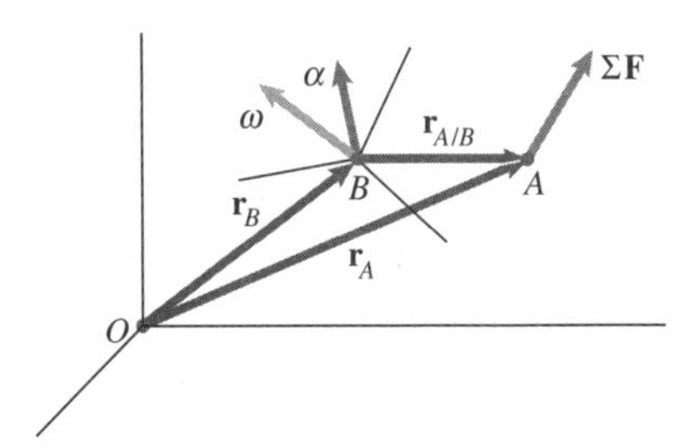

An inertial reference frame (origin O) and a reference frame undergoing an arbitrary motion (origin B).

We can analyze an object's motion relative to the reference frame undergoing the arbitrary motion (secondary reference frame e.g., a reference frame fixed with respect to a moving vehicle) by writing Newton's second law for an object A of mass m. This lead to the equation

$$\sum \mathbf{F} - m\mathbf{a}_B + 2\omega \times (\mathbf{v}_A)_{\text{rel}} + \alpha \times \mathbf{r}_{A/B} + \omega \times \left(\omega \times \mathbf{r}_{A/B}\right) = m\,(\mathbf{a}_A)_{\text{rel}}$$

from which we can find $(\mathbf{a}_A)_{\text{rel}}$ given the forces acting on A (i.e., $\sum \mathbf{F}$) and the secondary reference frame's motion ($\mathbf{a}_B$, α and ω).

Helpful Tips and Suggestions

- When using the relative velocity and acceleration equations

$$\mathbf{v}_A = \mathbf{v}_B + \omega \times \mathbf{r}_{A/B}$$
$$\mathbf{a}_A = \mathbf{a}_B + \alpha \times \mathbf{r}_{A/B} - \omega^2 \mathbf{r}_{A/B}$$

  the choice of point B is essential. It should always be a point whose velocity/acceleration is *known* or *easy to find*.
- Recall that the instantaneous center of zero velocity *does not, in general, have zero acceleration* so that although $\mathbf{v}_C = \mathbf{0}$, $\mathbf{a}_C \neq \mathbf{0}$.

Review Questions

1. What is the definition of a rigid body?
2. What is two-dimensional or planar motion of a rigid body?
3. How is the angular velocity vector of a rigid body defined?
4. If you know the direction of the angular velocity vector of a rigid body, how do you determine the direction of rotation of the body?
5. If you know the angular velocity vector and the velocity of one point of a rigid body, how can you determine the velocity of a different point of the rigid body?
6. If a point on a rigid body is an instantaneous center, what is its velocity?
7. For a rigid body in plane motion, is the instantaneous center always located either on the body or at a finite distance from the body?
8. Once the instantaneous center (C) has been found we can write

$$\begin{aligned}\mathbf{v}_A &= \mathbf{v}_C + \omega \times \mathbf{r}_{A/C} \\ &= \omega \times \mathbf{r}_{A/C}.\end{aligned}$$

   Is it the case that we can also write

$$\mathbf{a}_A = \alpha \times \mathbf{r}_{A/C} - \omega^2 \mathbf{r}_{A/C} \ ?$$

9. What does it mean when the term $(\mathbf{v}_A)_{\mathrm{rel}}$ is zero?
10. What is the definition of the angular acceleration vector?
11. What is meant by a body-fixed reference frame?
12. What is the definition of the term $(\mathbf{a}_A)_{\mathrm{rel}}$ in Eq. (17.4)?

18

Planar Dynamics of Rigid Bodies

Main Goals of this Chapter:

- To develop the planar equations of motion for a rigid body.
- To discuss applications of these equations to bodies undergoing translation, rotation about a fixed axis, and general planar motion.

18.1 Preview of the Equations of Motion

- The equations of motion of a rigid body for general planar motion (including both translational and angular motion) are given by

$$\sum \mathbf{F} = m\mathbf{a} \quad \text{(translational equation)}, \tag{18.1}$$

$$\sum M = I\alpha \quad \text{(angular motion)}, \tag{18.2}$$

  where $\sum \mathbf{F}$ is the sum of the external forces acting on the body, m is the mass of the body, $\mathbf{a}$ is the acceleration of the body's center of mass, $\sum M$ is the sum of the external moments about the body's center of mass, I is the body's moment of inertia about its center of mass and α represents the magnitude of the body's angular acceleration.
  Note. If the rigid body *rotates* about a fixed axis O, Eq. (18.2) can be replaced by

$$\sum M_O = I_O\alpha,$$

  where I_O is the body's moment of inertia about O.
- If the external forces and couples acting on a rigid body in planar motion are known, we can use these equations to determine *the acceleration of the center of mass of the body and its angular acceleration.*

18.2 Momentum Principles for a System of Particles

Force-Linear Momentum Principle

- The total external force on a system of particles equals the rate of change of the product of the total mass of the system and the velocity of its center of mass:

$$\sum \mathbf{F} = \frac{d}{dt}(m\mathbf{v}).$$

- *Since any object or collection of objects, including a rigid body, can be modeled as a system of particles,* this result is one of the most general and elegant in all mechanics.
- Further, if the total mass m is constant, we obtain

$$\sum \mathbf{F} = m\mathbf{a},$$

where $\mathbf{a} = \dfrac{d\mathbf{v}}{dt}$ is the acceleration of the body's mass center.

Moment-Angular-Momentum Principles

- Let $\mathbf{r}_i$ be the position of the ith particle of a system of particles relative to a fixed point O, $\mathbf{r}$ the position of the center of mass of the system, and $\mathbf{R}_\text{i}$ the position of the ith particle relative to the center of mass. The sum of the moments due to external forces about O is equal to the *rate of change of the total angular momentum about O*:

$$\sum \mathbf{M}_O = \frac{d\mathbf{H}_O}{dt}.$$

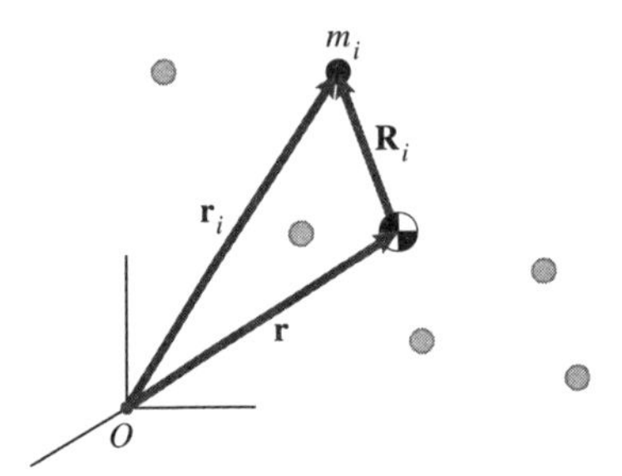

The vector R_i is the position vector of the ith particle relative to the center of mass.

- The *angular momentum* about O is

$$\mathbf{H}_O = \sum_i \mathbf{r}_i \times m_i \mathbf{v}_i$$

where m_i is the mass of the ith particle and $\mathbf{v}_i$ is its velocity. This relationship can also be written as

$$\sum \mathbf{M}_O = \frac{d}{dt}(\mathbf{r} \times m\mathbf{v} + \mathbf{H})$$

where m is the total mass of the system of particles, $\mathbf{v}$ is the velocity of the center of mass, and

$$\mathbf{H} = \sum_i \mathbf{R}_i \times m_i \frac{d\mathbf{R}_i}{dt}$$

is the *total angular momentum* about the center of mass.

- *The sum of the moments due to external forces about the center of mass is equal to the rate of change of the total angular momentum about the center of mass*

$$\sum \mathbf{M} = \frac{d\mathbf{H}}{dt}.$$

- The angular momenta about point O and about the center of mass are related by

$$\mathbf{H}_O = \mathbf{H} + \mathbf{r} \times m\mathbf{v}.$$

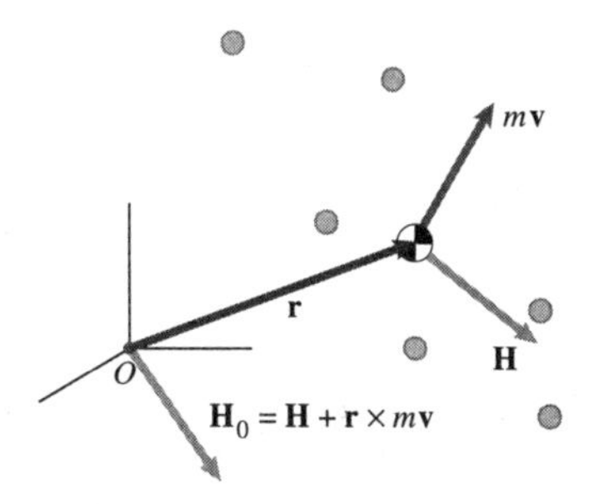

The angular momentum about O equals the sum of the angular momentum about the center of mass and the angular momentum about O due to the velocity of the center of mass.

18.3 Derivation of the Equations of Motion

- The equations of motion for a rigid body in planar motion include Newton's second law

$$\sum \mathbf{F} = m\mathbf{a}, \tag{18.3}$$

where $\mathbf{a}$ is the acceleration of the body's mass center. This equation is coupled with an equation of angular motion as follows:

Rotation about a Fixed Axis

- Let O be a point that is stationary relative to an inertial reference frame. *If the rigid body rotates about O*, the total moment about O equals the product of the moment of inertia about O and the angular acceleration

$$\sum M_O = I_O \alpha. \tag{18.4}$$

General Planar Motion

- *In general planar motion,* the total moment about the center of mass equals the product of the moment of inertia about the center of mass and the angular acceleration

$$\sum M = I\alpha. \tag{18.5}$$

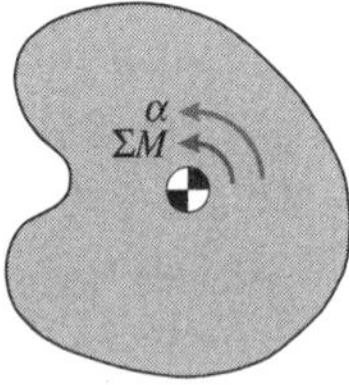

18.4 Applications

The Eqs. (18.3) and (18.5) (Eq. (18.4) which is a special case, and more convenient version, of Eq. (18.5) when the motion is *pure rotation* about a fixed point O) can be used to obtain information about an object's motion, or to determine the values of unknown forces or couples acting on the object, or both. Doing so involves the following procedure:

Procedure for Solving Problems

- **Draw the Free-Body Diagram**. Isolate the object and identify the external forces and couples acting on it.
- **Apply the Equations of Motion.** Write equations of motion suitable for the type of motion, choosing an appropriate coordinate system for applying Newton's second law. For example, if the center of mass moves in a circular path, it may be advantageous to use normal and tangential components.
- **Determine Kinematic Relationships.** If necessary, supplement the equations of motion with relationships between the acceleration of the center of mass and the angular acceleration of the object.

Note. *The best approach depends, in part, on the type of motion involved.*

Translation

If a rigid body is in translation, *only* Newton's second law (Eq. (18.3)) is required to determine its motion. There is no rotational motion to determine. Nevertheless, it may be necessary to apply the angular equation of motion to determine unknown forces or couples. Since $\alpha = \mathbf{0}$, Eq. (18.5) states that the total moment *about the center of mass* equals zero:

$$\sum M = 0.$$

A rigid body in translation. There is no rotational motion to determine.

Rotation about a Fixed Axis

In the case of rotation about a fixed axis, *only* Eq. (18.4) is required to determine the rotational motion, although it may be necessary to use Newton's second law (Eq. (18.3)) to determine unknown forces or couples.

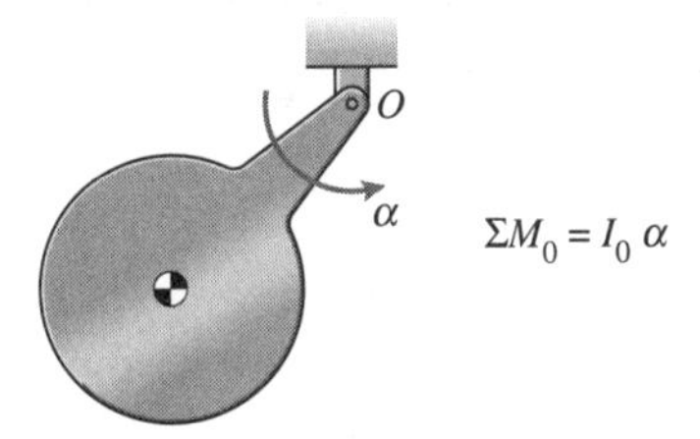

A rigid body rotating about O. You need only the equation of angular motion about O to determine the angular acceleration of the body.

General Planar Motion

If a rigid body undergoes general planar motion, *both* Newton's second law (Eq. (18.3)) and the equation of angular motion (Eq. (18.5)) are required to determine its motion. If the motion of the center of mass and the rotational motion are not independent (e.g., in a rolling disk), there will be more unknown quantities than equations of motion. In such cases, it is necessary to obtain additional equations by deriving kinematic relationships between the acceleration of the center of mass and the angular acceleration.

18.5 Numerical Solutions

- When the forces and couples acting on a rigid body are known, the equations of motion can be used to determine the acceleration of the center of mass and the angular acceleration of the body.
- In some situations, we can then integrate the equations to obtain closed-form expressions for the velocity and position of the center of mass and for the angular velocity and angular position as functions of time. However, if the functions describing the accelerations are too complicated, or the forces and couples are known in terms of continuous or analog data instead of equations, a numerical method must be used to determine the velocities and positions as functions of time.
- In fact, the angular position and angular velocity can be determined using the same numerical procedure (finite-difference method) used for determining the position and velocity of the center of mass of an object as functions of time. (See Chapter 14.)

APPENDIX - Moments of Inertia

- **Definition.** The moment of inertia of an object about an axis L_O is

$$I_O = \int_m r^2 dm,$$

where r is the perpendicular distance from L_O to the differential element of mass dm.

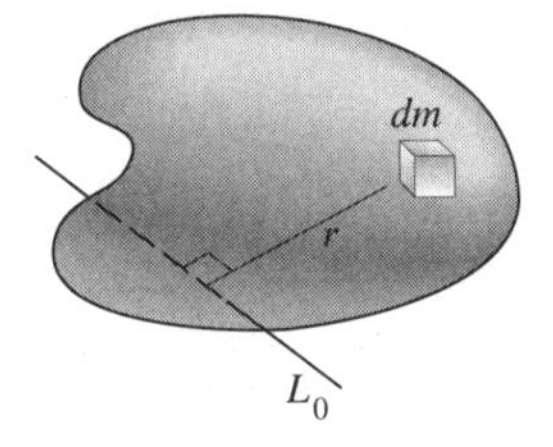

- **Parallel-Axis Theorem.** Let L be an axis *through the center of mass of an object*, and let L_O be a parallel axis. The moment of inertia I_O, about L_O, is given in terms of the moment of inertia I, about L, by the parallel-axis theorem

$$I_O = I + d^2 m,$$

where m is the mass of the object and d is the distance between L and L_O.

Helpful Tips and Suggestions

- Remember always to draw a *free-body diagram* to account for all the external forces and couples that act on the body.
- Before attempting any of the problems, study the Examples in the text. Re-work a few of these *yourself* knowing you have the full solution available. This is an excellent way to reinforce ideas and understand the relevant material.

Review Questions

1. Why can results for systems of particles be used for a rigid body?
2. What are the equations of motion for general planar motion of a rigid body?
3. What is the procedure for solving problems involving general planar motion of a rigid body?
4. If a rigid body is in translation, what are the equations which need to be solved to determine the motion?
5. If a rigid body is in rotation about a fixed axis, what are the equations which need to be solved to determine the motion?
6. When can the equation $\sum M_O = I_O \alpha$. (Eq.(18.4)) be used in place of $\sum M = I\alpha$ Eq. (18.5)?
7. How many *scalar* equations of motion are there in the description of general planar motion of a rigid body?
8. How many *scalar* equations of motion would you expect to describe the three-dimensional motion of a rigid body?

19

Energy and Momentum Methods in Rigid-Body Dynamics

Main Goals of this Chapter:

- To develop work and energy methods for the solution of problems involving *planar motions of rigid bodies.*
- To develop impulse and momentum methods for the solution of problems involving *planar motions of rigid bodies.*

19.1 Principle of Work and Energy

- **Principle of Work and Energy for a Rigid Body.**
  The work done by external forces and couples as a rigid body moves between two positions equals the change in the total kinetic energy of the body:

$$U_{12} = T_2 - T_1.$$

- **Principle of Work and Energy for a System of Rigid Bodies.**
  The work done by external and internal forces and couples as a system of rigid bodies moves between two positions equals the change in the total kinetic energy of the system.

19.2 Kinetic Energy

General Planar Motion

- The kinetic energy of a rigid body in general planar motion is

$$T = \frac{1}{2}mv^2 + \frac{1}{2}I\omega^2, \qquad (19.1)$$

where v is the magnitude of the velocity of the center of mass of the body and I is the moment of inertia about the center of mass.

Fixed-Axis Rotation

- A rigid body rotating about a fixed axis O is in general planar motion. Hence, it's kinetic energy is described by Eq. (19.1). However, its kinetic energy can also be expressed as

$$T = \frac{1}{2} I_O \omega^2,$$

where ω is the magnitude of the body's angular velocity about O and I_O is the body's moment of inertia about O.

19.3 Work and Potential Energy

- The *work done* on a rigid body by a force $\mathbf{F}$ is

$$U_{12} = \int_{(\mathbf{r}_P)_1}^{(\mathbf{r}_P)_2} \mathbf{F} \cdot d\mathbf{r}_P,$$

where $\mathbf{r}_P$ is the position of the point of application of $\mathbf{F}$.

- If the point of application is stationary, or if its direction of motion is perpendicular to $\mathbf{F}$, no work is done.
- A force $\mathbf{F}$ is conservative if a potential energy V exists such that

$$\mathbf{F} \cdot d\mathbf{r}_P = -dV.$$

- In terms of its potential energy, the work done by a conservative force $\mathbf{F}$ is

$$U_{12} = -(V_2 - V_1)$$

where V_1 and V_2 are the values of V at $(\mathbf{r}_P)_1$ and $(\mathbf{r}_P)_2$.

- The *work done* by a couple M on a rigid body in planar motion as the body rotates from θ_1 to θ_2 in the direction of M is

$$U_{12} = \int_{\theta_1}^{\theta_2} M \, d\theta.$$

- If M is *constant*, the work done is

$$U_{12} = M(\theta_2 - \theta_1) \quad \text{(constant couple)}.$$

- A couple M is *conservative* if a potential energy V exists such that

$$M \, d\theta = -dV.$$

- The *work done by a conservative couple* is

$$U_{12} = -(V_2 - V_1).$$

- The potential energy of a linear torsional spring that exerts a couple $k\theta$ in the direction opposite to angular displacement θ of the spring is

$$V = \frac{1}{2}k\theta^2.$$

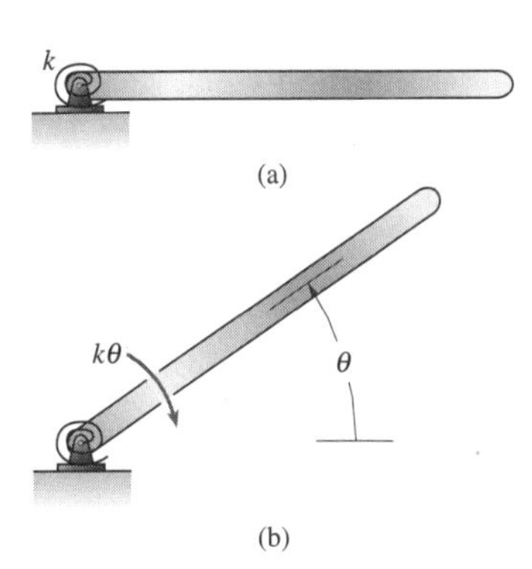

(a) A linear torsional spring connected to a bar. (b) The spring exerts a couple of magnitude $k\theta$ in the direction opposite that of the bar's rotation.

- **Conservation of Energy for a Rigid Body.**
  If all the forces and couples that do work on a rigid body are conservative, the sum of the kinetic energy and the total potential energy of the body is constant:

$$T + V = \text{constant}.$$

Solving Problems Using Work and Energy

1. *Identify the forces and couples that do work.* Use free-body diagrams to determine which external forces and couples do work.
2. *Apply the principle of work and energy or conservation of energy.* Either equate the total work done during a change in position to the change in the kinetic energy, or equate the sum of the kinetic and potential energies at two positions.
3. *Determine the kinematic relationships.* To complete the solution, it will often be necessary to obtain relations between velocities of points of rigid bodies and their angular velocities.

19.4 Power

- The power (rate at which work is done on the rigid body) transmitted to a rigid body *by a force* **F** is

$$P = \mathbf{F} \cdot \mathbf{v}_P$$

  where $\mathbf{v}_P$ is the velocity of the point of application of **F**.
- The power transmitted to a rigid body in planar motion *by a couple M* is

$$P = M\omega.$$

- The *average power* transferred to a rigid body during an interval of time is equal to the change in kinetic energy of the body, or the total work done during that time, divided by the interval of time:

$$P_{\text{av}} = \frac{T_2 - T_1}{t_2 - t_1} = \frac{U_{12}}{t_2 - t_1}.$$

19.5 Principles of Impulse and Momentum

Linear Momentum

- *The principle of linear impulse and momentum:*

The linear impulse applied to a rigid body during an interval of time is equal to the change in linear momentum of the body during that time

$$\int_{t_1}^{t_2} \sum \mathbf{F} dt = m\mathbf{v}_2 - m\mathbf{v}_1.$$

— This result can also be expressed in terms of the average of the total force with respect to time:

$$(t_2 - t_1) \sum \mathbf{F}_{\text{av}} = m\mathbf{v}_2 - m\mathbf{v}_1.$$

- **Conservation of Linear Momentum**. If the only forces acting on two rigid bodies A and B are the forces they exert on each other, or if the forces are negligible, then the total linear momentum of the bodies is conserved:

$$m_A \mathbf{v}_B + m_B \mathbf{v}_B = \text{constant}.$$

Angular Momentum

- The angular momentum about the center of mass of a rigid body in planar motion is

$$H = I\omega,$$

where I is the moment of inertia of the body about its center of mass.

- **Principle of Angular Impulse and Momentum—First Form.**

The angular impulse about the center of mass during the interval from t_1 to t_2 is equal to the change in the body's angular momentum about its center of mass during that interval.

$$\int_{t_1}^{t_2} \sum M dt = H_2 - H_1. \tag{19.2}$$

- The angular momentum of a rigid body about a fixed point O is

$$H_O = (\mathbf{r} \times m\mathbf{v}) \cdot \mathbf{k} + I\omega, \tag{19.3}$$

where $\mathbf{r}$ is the position of the center of mass relative to O and $\mathbf{v}$ is the velocity of the center of mass.

- **Principle of Angular Impulse and Momentum—Second Form.**

The angular impulse about O during the interval from t_1 to t_2 is equal to the change in the rigid body's angular momentum about O during that interval.

$$\int_{t_1}^{t_2} \sum M_O dt = H_{O2} - H_{O1}.$$

- In terms of the averages of the total moments with respect to time, Eqs. (19.2)–(19.3) are

$$(t_2 - t_1) \sum M_{\text{av}} = H_2 - H_1,$$

$$(t_2 - t_1) \left(\sum M_O\right)_{\text{av}} = H_{O2} - H_{O1}.$$

- **Conservation of Angular Momentum.** If two rigid bodies A and B in planar motion are subjected only to internal forces and couples, or if the total moment due to external forces and couples about a fixed point O is zero, the total angular momentum of A and B about O is *conserved*:

$$H_{OA} - H_{OB} = \text{constant}.$$

19.6 Impacts

Conservation of Momentum

Suppose that two rigid bodies A and B in two-dimensional motion in the same plane *collide*.

- **Linear Momentum.** If other forces are negligible in comparison to the impact forces that A and B exert on each other, then the total linear momentum of A and B is *conserved.*

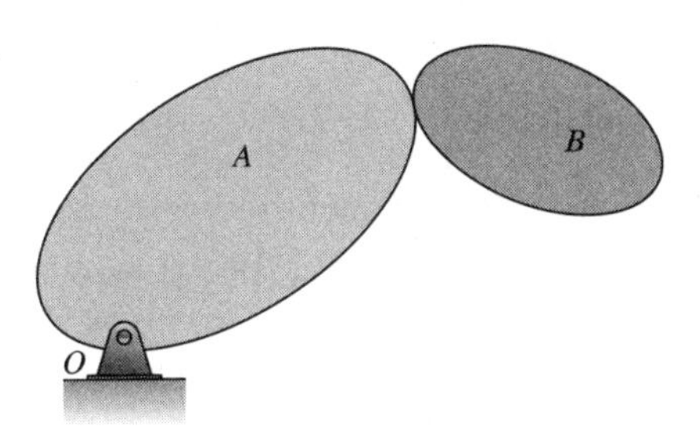

Rigid bodies A and B colliding. Because of the pin support, their total linear momentum is *not* conserved, but their total angular momentum about O is conserved.

- **Angular Momentum.** If other forces and couples are negligible in comparison to the impact forces and couples that A and B exert on each other, then the total angular momentum of A and B about any fixed point O *is conserved.*
- If, in addition, A and B exert only forces on each other at their point of impact, P, the angular momentum about P of *each* rigid body is *conserved.*
- If one of the rigid bodies A and B has a pin support at point O , the *total* angular momentum of both bodies about that point is conserved.

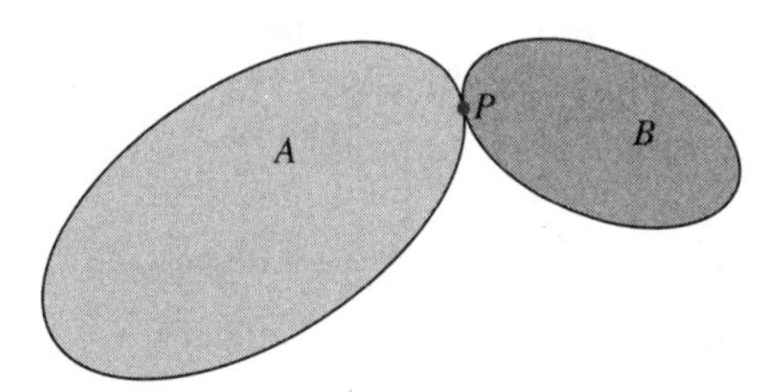

Rigid bodies A and B colliding at P. If forces are exerted only at P, the angular momentum of A about P and the angular momentum of B about P are each conserved.

Coefficient of Restitution

- Let P be the point of impact. The normal components of the velocities *at* P are related to the coefficient of restitution e by

$$e = \frac{(\mathbf{v}'_{BP})_x - (\mathbf{v}'_{AP})_x}{(\mathbf{v}_{AP})_x - (\mathbf{v}_{BP})_x}$$

Note. In arriving at this equation, we assumed that the contacting surfaces were smooth, so *the collision exerts no force on A or B in the direction tangential to their contacting surfaces* (frictional forces resulting from impact are negligible).

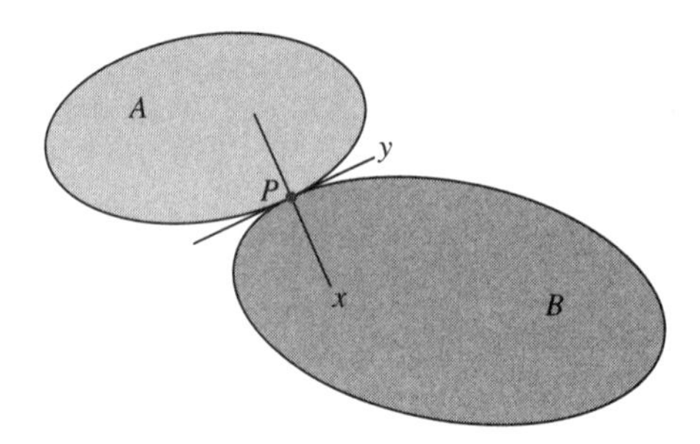

Rigid bodies A and B colliding at P. The x axis is perpendicular to the contacting surfaces.

Helpful Tips and Suggestions

- Remember that unlike force, acceleration, or displacement, energy is a *scalar.*
- A brief review of Chapter 15 may prove helpful in solving problems involving energy since computations for kinetic energy require a kinematic analysis of velocity.
- Only problems involving *conservative forces* (weights and springs) may be solved using *conservation of energy.* Friction or other drag-resistant forces, which depend on velocity or acceleration are nonconservative (a portion of the work done by such forces is transformed into thermal energy which dissipates into the surroundings and may not be recovered). When such forces enter into the problem, use the *principle of work and energy.*
- Review Chapter 16 before solving problems involving momentum methods.

Review Questions

1. What is the principle of work and energy for a rigid body?
2. What is the kinetic energy of a rigid body in general planar motion?
3. If all of the forces and couples that do work on a rigid body are conservative, what can you say about the sum of the kinetic energy of the rigid body and the total potential energy?
4. If the total moment about a fixed point O due to the external forces and couples acting on a rigid body is zero during an interval of time, what can you say about the rigid body's angular momentum about O?
5. When two rigid bodies collide, you cannot always assume that their total linear momentum is conserved. Explain why.
6. In the figure, is total linear momentum conserved?
7. In the figure, is total angular momentum conserved about point O?

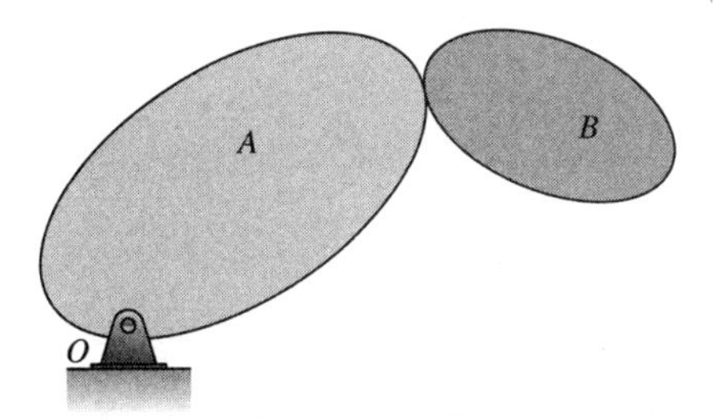

Rigid bodies A and B colliding. Because of the pin support, their total linear momentum is *not* conserved, but their total angular momentum about O is conserved.

8. Is the coefficient of restitution for two colliding rigid bodies defined in terms of their centers of mass or in terms of their velocities at the point of impact?

20

Three-Dimensional Kinematics and Dynamics of Rigid Bodies

Main Goals of this Chapter:

- To derive the three-dimensional equations of motion for a rigid body and use them to analyze simple three-dimensional motions.
- To introduce Euler angles for the description of the orientation of a rigid body in three dimensions.
- To express the equations of angular motion in terms of Euler angles

20.1 Kinematics

Velocities and Accelerations

- The *velocity* of a point A of a rigid body relative to a given reference frame is given in terms of the velocity of a point B of the rigid body and the rigid body's angular velocity by

$$\mathbf{v}_A = \mathbf{v}_B + \omega \times \mathbf{r}_{A/B}.$$

- The *acceleration* of point A is given in terms of the acceleration of point B, the rigid body's angular acceleration, and its angular velocity by

$$\mathbf{a}_A = \mathbf{a}_B + \alpha \times \mathbf{r}_{A/B} + \omega \times (\omega \times \mathbf{r}_{A/B}).$$

Moving Reference Frames

Consider a secondary reference frame xyz with angular velocity Ω relative to a primary reference frame, and a rigid body with angular velocity ω relative to the primary reference frame.

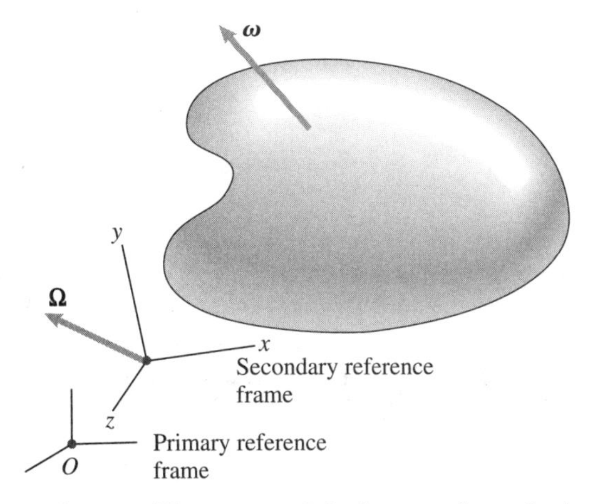

The primary and secondary reference frames. The vector Ω is the angular velocity of the secondary reference frame relative to the primary reference frame. The vector ω is the angular velocity of the rigid body relative to the primary reference frame.

- If the secondary reference frame is *body fixed*,

$$\Omega = \omega.$$

  — The rigid body's *angular acceleration* relative to the primary reference frame is

$$\alpha = \frac{d\omega_x}{dt}\mathbf{i} + \frac{d\omega_y}{dt}\mathbf{j} + \frac{d\omega_z}{dt}\mathbf{k} + \Omega \times \omega$$

- If the secondary reference frame is *not body fixed,* it is often convenient to express ω as the sum of Ω and the angular velocity ω_{rel} of the rigid body relative to the secondary reference frame:

$$\omega = \Omega + \omega_{\text{rel}}.$$

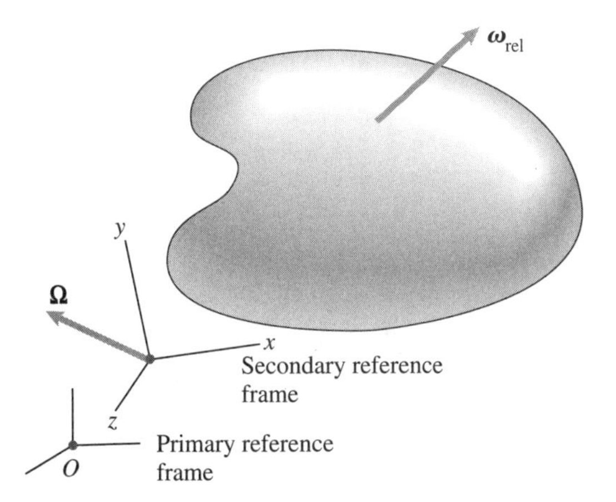

The vector ω_{rel} is the rigid body's angular velocity relative to the secondary reference frame, and the vector ω is the angular velocity of the secondary reference frame relative to the primary reference frame. The rigid body's angular velocity relative to the primary reference frame is $\omega_{\text{rel}} + \omega$.

20.2 Euler's Equations

The three-dimensional equations of motion for a rigid body are called *Euler's Equations*. They include, in addition to equations of *angular motion*, Newton's second law:

$$\sum \mathbf{F} = m\mathbf{a}.$$

In deriving the *equations of angular motion*, we consider first the special case of rotation about a fixed point and then general three-dimensional motion.

Rotation About a Fixed Point

- For a rigid body *rotating about a fixed point* O, the equations of angular motion are expressed in terms of the components of the total moment about O as

$$\begin{aligned}\sum M_{Ox} = {} & I_{xx}\frac{d\omega_x}{dt} - I_{xy}\frac{d\omega_y}{dt} - I_{xz}\frac{d\omega_z}{dt} \\ & -\Omega_z\left(-I_{yx}\omega_x + I_{yy}\omega_y - I_{yz}\omega_z\right) \\ & +\Omega_y\left(-I_{zx}\omega_x - I_{zy}\omega_y + I_{zz}\omega_z\right),\end{aligned}$$

$$\begin{aligned}\sum M_{Oy} = {} & -I_{yx}\frac{d\omega_x}{dt} + I_{yy}\frac{d\omega_y}{dt} - I_{yz}\frac{d\omega_z}{dt} \\ & +\Omega_z\left(I_{xx}\omega_x - I_{xy}\omega_y - I_{xz}\omega_z\right) \\ & -\Omega_x\left(-I_{zx}\omega_x - I_{zy}\omega_y + I_{zz}\omega_z\right),\end{aligned} \tag{20.1}$$

and

$$\begin{aligned}\sum M_{Oz} = {} & -I_{zx}\frac{d\omega_x}{dt} - I_{zy}\frac{d\omega_y}{dt} + I_{zz}\frac{d\omega_z}{dt} \\ & -\Omega_y\left(I_{xx}\omega_x - I_{xy}\omega_y - I_{xz}\omega_z\right) \\ & +\Omega_x\left(-I_{yx}\omega_x + I_{yy}\omega_y - I_{yz}\omega_z\right),\end{aligned}$$

where ω is the angular velocity of the rigid body and Ω is the angular velocity of the chosen secondary reference frame with origin at O.

♦ If the secondary reference frame is *body fixed*,

$$\Omega = \omega.$$

Note. In writing the equations (20.1), we have assumed that the moments and products of inertia are constants. This is so when the secondary reference frame is body fixed, but must be confirmed when it is not.

♦ The equations (20.1) can also be written in matrix form, which introduces the *inertia matrix* of the rigid body, given by

$$[I] = \begin{bmatrix} I_{xx} & -I_{xy} & -I_{xz} \\ -I_{yx} & I_{yy} & -I_{yz} \\ -I_{zx} & -I_{zy} & I_{zz} \end{bmatrix}$$
$$= \begin{bmatrix} \int_m \left(y^2+z^2\right) dm & -\int_m xy dm & -\int_m xz dm \\ -\int_m yx dm & \int_m \left(x^2+z^2\right) dm & -\int_m yz dm \\ -\int_m zx dm & -\int_m zy dm & \int_m \left(x^2+y^2\right) dm \end{bmatrix},$$

where x, y, and z are the coordinates of the differential element of mass dm. The terms I_{xx}, I_{yy}, and I_{zz} are the *moments of inertia* about the x, y, and z axes, respectively, and I_{xy}, I_{yz}, and I_{zx} are the *products of inertia*.

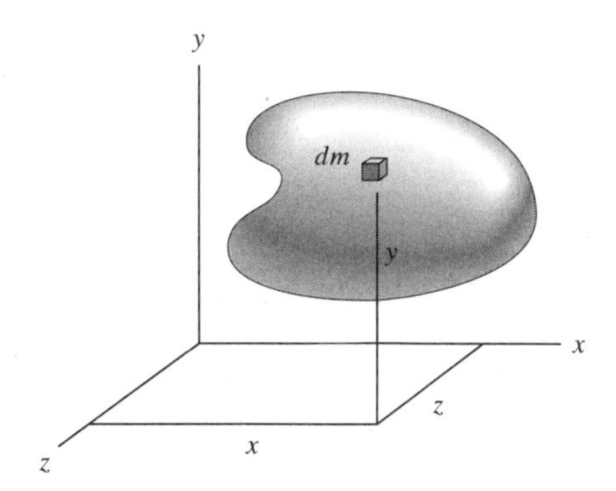

Determining the moments and products of inertia by modeling an object as a continuous distribution of mass.

General Three-Dimensional Motion

In the case of general three-dimensional motion, the equations of angular motion are identical in form to those for motion about a fixed point, except that they are expressed in terms of the components of the total moment about the center of mass and the origin of the secondary reference frame is at the center of mass.

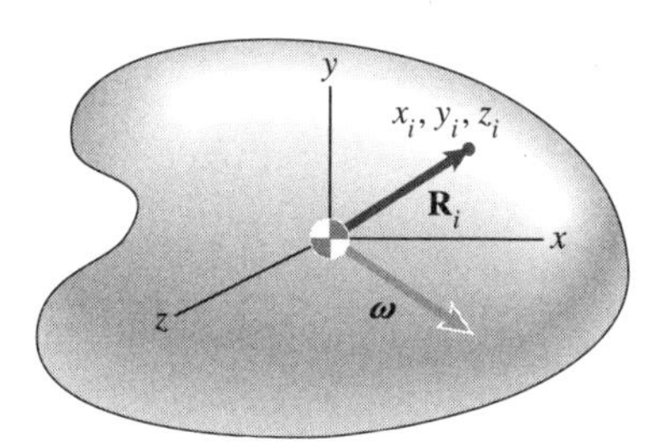

Coordinate system with its origin at the center of mass of the body.

In other words,

$$\begin{aligned}\sum M_x &= I_{xx}\frac{d\omega_x}{dt} - I_{xy}\frac{d\omega_y}{dt} - I_{xz}\frac{d\omega_z}{dt}\\ &\quad -\Omega_z\left(-I_{yx}\omega_x + I_{yy}\omega_y - I_{yz}\omega_z\right)\\ &\quad +\Omega_y\left(-I_{zx}\omega_x - I_{zy}\omega_y + I_{zz}\omega_z\right),\end{aligned}$$

$$\begin{aligned}\sum M_y &= -I_{yx}\frac{d\omega_x}{dt} + I_{yy}\frac{d\omega_y}{dt} - I_{yz}\frac{d\omega_z}{dt}\\ &\quad +\Omega_z\left(I_{xx}\omega_x - I_{xy}\omega_y - I_{xz}\omega_z\right)\\ &\quad -\Omega_x\left(-I_{zx}\omega_x - I_{zy}\omega_y + I_{zz}\omega_z\right),\end{aligned} \qquad (20.2)$$

and

$$\sum M_z = -I_{zx}\frac{d\omega_x}{dt} - I_{zy}\frac{d\omega_y}{dt} + I_{zz}\frac{d\omega_z}{dt}$$
$$-\Omega_y\left(I_{xx}\omega_x - I_{xy}\omega_y - I_{xz}\omega_z\right)$$
$$+\Omega_x\left(-I_{yx}\omega_x + I_{yy}\omega_y - I_{yz}\omega_z\right),$$

Procedure for Using Euler Equations to Analyze Three-Dimensional Motions of Rigid Bodies

1. *Choose a coordinate system.* If an object rotates about a fixed point O, it is usually preferable to use a secondary coordinate system with its origin at O and express the equations of angular motion in the forms given by Eqs. (20.1). Otherwise, it is necessary to use a coordinate system with its origin at the center of mass of the body and express the equations of angular motion in the forms given by Eqs. (20.2). In either case, it is usually preferable to choose a coordinate system that simplifies the determination of the moments and products of inertia.
2. *Draw the free-body diagram.* Isolate the object and identify the external forces and couples acting on it.
3. *Apply the equations of motion.* Use Newton's second law and the equations of angular motion to relate the forces and couples acting on the object to the acceleration of its center of mass and its angular acceleration.

Equations of Planar Motion

- The equations of angular motion for a rigid body in *planar motion*, that is

$$\sum M_O = I_O\frac{d\omega}{dt} \quad \text{(rotation about a fixed axis)},$$
$$\sum M = I\frac{d\omega}{dt} \quad \text{(general planar motion)}$$

can be obtained from the three-dimensional equations (Eqs (20.1) or (20.2)) as *special cases* under the assumptions of planar motion.

20.3 The Euler Angles

Euler angles are used to specify the orientation of a rigid body in three-dimensions.

Objects with an Axis of Symmetry

- In the case of an object with an axis of rotational symmetry, the orientation of the xyz system relative to the reference XYZ system is specified by the *precession angle* ψ and the *nutation angle* θ. The rotation of the object relative to the xyz system is specified by the *spin angle* ϕ.

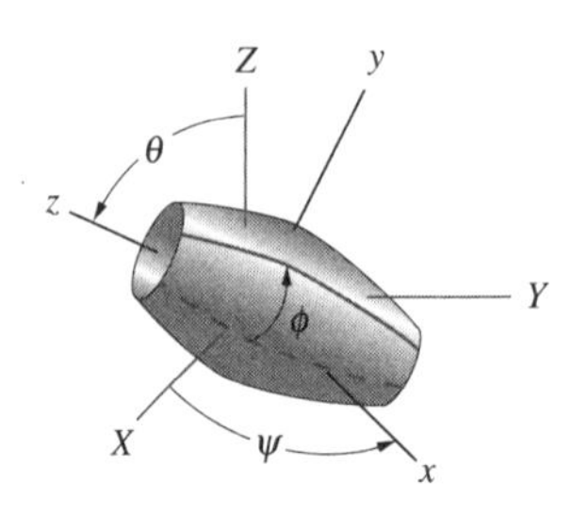

The rotation ϕ of the object relative to the xyz system.

- The components of the rigid body's angular velocity relative to the *XYZ* system is given by

$$\omega_x = \dot{\theta}, \quad \omega_y = \dot{\psi}\sin\theta \quad \text{and} \quad \omega_z = \dot{\phi} + \dot{\psi}\cos\theta.$$

- The equations of angular motion, expressed in *terms of the Euler angles*, are

$$\sum M_x = I_{xx}\ddot{\theta} + (I_{zz} - I_{xx})\,\dot{\psi}^2 \sin\theta\cos\theta + I_{zz}\dot{\phi}\dot{\psi}\sin\theta,$$

$$\sum M_y = I_{xx}(\ddot{\psi}\sin\theta + 2\dot{\psi}\dot{\theta}\cos\theta) - I_{zz}(\dot{\phi}\dot{\theta} + \dot{\psi}\dot{\theta}\cos\theta),$$

$$\sum M_z = I_{zz}(\ddot{\phi} + \ddot{\psi}\cos\theta - \dot{\psi}\dot{\theta}\sin\theta).$$

- In the *steady-precession* of an axisymmetric spinning object, the spin rate $\dot{\phi}$, the nutation angle θ, and the precession rate $\dot{\psi}$ are *assumed to be constant*. With these assumptions, the equations of angular motion reduce to

$$\sum M_x = (I_{zz} - I_{xx})\,\dot{\psi}^2 \sin\theta\cos\theta + I_{zz}\dot{\phi}\dot{\psi}\sin\theta,$$

$$\sum M_y = 0,$$

$$\sum M_z = 0.$$

Arbitrary Objects

- In the case of an arbitrary object, the moments and products of inertia will be constants only if the *xyz* coordinate system is body fixed. This means that *three Euler angles* are needed to specify the orientation of the coordinate system, and the resulting equations of angular motion are *more complicated.* In fact, assuming that the body-fixed coordinate system *xyz* is a set of principal axes, Eqs. (20.2) become

$$\begin{aligned}\sum M_x = {} & I_{xx}\ddot{\psi}\sin\theta\sin\phi + I_{xx}\ddot{\theta}\cos\phi \\ & + I_{xx}\dot{\psi}\dot{\theta}\cos\theta\sin\phi + \dot{\psi}\dot{\phi}\cos\phi\sin\theta - \dot{\phi}\dot{\theta}\sin\phi \\ & - (I_{xx} - I_{zz})\left(\dot{\psi}\sin\theta\cos\phi - \dot{\theta}\sin\phi\right)\left(\dot{\psi}\cos\theta + \dot{\phi}\right),\end{aligned}$$

$$\begin{aligned}\sum M_y = {} & I_{yy}\ddot{\psi}\sin\theta\cos\phi - I_{yy}\ddot{\theta}\sin\phi \\ & + I_{yy}(\dot{\psi}\dot{\theta}\cos\theta\cos\phi - \dot{\psi}\dot{\phi}\sin\phi\sin\theta - \dot{\phi}\dot{\theta}\cos\phi \\ & - (I_{zz} - I_{xx})\left(\dot{\psi}\cos\theta + \dot{\phi}\right)\left(\dot{\psi}\sin\theta\sin\phi + \dot{\theta}\cos\phi\right),\end{aligned} \qquad (20.3)$$

and

$$\begin{aligned}\sum M_z = {} & I_{zz}\ddot{\psi}\cos\theta + I_{zz}\ddot{\phi} - I_{zz}\dot{\psi}\dot{\theta}\sin\theta \\ & - \left(I_{xx} - I_{yy}\right)\left(\dot{\psi}\sin\theta\cos\phi - \dot{\theta}\sin\phi\right)\left(\dot{\psi}\sin\theta\sin\phi + \dot{\theta}\cos\phi\right).\end{aligned}$$

- If the Euler angles and their first and second derivatives with respect to time are known, are known, Eqs. (20.3) can be solved for the components of the total moment.
- Alternatively, if the total moment, the Euler angles, and the first derivatives of the Euler angles are known, Eqs. (20.3) can be solved for the second derivatives of the Euler angles. In this way, the Euler angles can be determined as a function of time when the total moment is known as a function of time, but numerical integration is usually necessary.

Helpful Tips and Suggestions

- The *vector* form of the kinematic equations for $\mathbf{v}_B$ and $\mathbf{a}_B$ (see Eqs. (20.1) and (20.2) in text) are identical in both plane and three-dimensional kinematics. This, in itself, is sufficient reason to become comfortable with vector calculus and vector algebra when studying problems in mechanics.
- Examples and worked problems are the key—especially in three-dimensional problems where it is much more difficult to visualize the motion. Work through the examples in the text *yourself* before attempting the problems. You will gain a much better understanding of the equations and how to apply them.

Review Questions

1. Suppose that the angular velocity of a secondary reference frame relative to a primary reference frame is Ω. If the angular velocity of a rigid body relative to the secondary reference frame is ω_{rel}, what is the rigid body's angular velocity relative to the primary reference frame?
2. What are Euler's equations?
3. How are the moments and products of inertia of a rigid body defined?
4. What are the differences between Eqs. (20.1) and (20.2)?
5. What's the only difference when using the equations

$$\mathbf{v}_B = \mathbf{v}_A + \omega \times \mathbf{r}_{B/A}$$

$$\mathbf{a}_B = \mathbf{a}_A + \alpha \times \mathbf{r}_{B/A} + \omega \times (\omega \times \mathbf{r}_{B/A})$$

in three-dimensional kinematics, as opposed to plane kinematics?

6. What are the Euler angles used for?
7. What is *steady precession* of an axisymmetric spinning object?

21

Vibrations

Main Goals of this Chapter:

- To discuss one-degree-of-freedom vibration of a rigid body subjected to conservative forces.
- To study the analysis of damped vibrations.
- To model and analyze the effect of forced vibrations.

Preliminaries

- A *vibration* is the periodic motion of a body or system of connected bodies displaced from a position of equilibrium.
  - *Free vibration*: when the motion is maintained by gravitational or elastic restoring forces; that is, the vibration is *free* from external periodic or intermittent forces applied to the system.
  - *Forced vibration*: caused by external periodic or intermittent forces applied to the system.
- *Undamped vibrations* can continue indefinitely since damping (e.g., frictional) effects are neglected in the analysis.
- *Damped vibrations* are vibrations which decay/die out with time due to the effects of internal and external frictional forces. In reality, the motion of all vibrating bodies is *actually damped*.
- If a vibrating system has *one degree of freedom*, it requires only *a single coordinate* (x) to specify completely the position of the system at any time t.

21.1 Conservative Systems

- The undamped, free vibrations of a system are due only to gravitational and elastic restoring forces acting on the system. Since these types of forces are conservative, it is also possible to use the *conservation of energy* equation to obtain the body's (natural) frequency or period of vibration.
- Small vibrations of many one-degree-of-freedom *conservative systems* relative to an equilibrium position are governed by the equation

$$\frac{d^2x}{dt^2} + \omega^2 x = 0 \ \textit{(undamped, free vibrations)} \tag{21.1}$$

where ω is a constant determined by the properties of the system.

- The general solution of this differential equation (21.1) is

$$x(t) = A \sin \omega t + B \cos \omega t$$

where A and B are two constants of integration generally determined from the initial conditions of the problem.
  — An alternative form of this general solution is

$$x(t) = E \sin(\omega t - \phi)$$

where again E and ϕ are new arbitrary constants to be determined. This type of motion (Eq. (21.1)) is referred to as *simple harmonic motion.*
We have the following vibrating characteristics of the system:
  — **Amplitude**: $E = \sqrt{A^2 + B^2}$ is amplitude of the vibration.
  — **Period:** $\tau = \dfrac{2\pi}{\omega}$ is time required to complete one complete oscillation or cycle
  — **Frequency:** The frequency is the number of cycles completed per unit of time. It is the reciprocal of the period:

$$f = \frac{1}{\tau} = \frac{\omega}{2\pi}.$$

The frequency is expressed in cycles/s i.e., $1Hz$ (Hertz) $= 1cycle/s = 2\pi rad/s$
- A system's period and frequency are determined by its physical properties, and do not depend on the functional form in which its motion is expressed.
- To obtain the equation for a one-degree-of-freedom conservative system *in the form (21.1)*, it may be necessary to linearize the system's equation of motion by assuming that the displacement from equilibrium is *small.*

21.2 Damped Vibrations

Frictional forces or damping mechanisms *damp out or attenuate* the vibration. In this section, we model the damping effect in vibrating systems by *adding* a 'damping effect' to the equation (21.1) (in which all damping was neglected)

- Small vibrations of many damped one-degree-of-freedom systems relative to an equilibrium position are governed by the equation

$$\frac{d^2x}{dt^2} + 2d\frac{dx}{dt} + \omega^2 x = 0, \tag{21.2}$$

♦ **Subcritical Damping.** If $d < \omega$, the system is said to be *subcritically damped.* In this case, the general solution of Eq. (21.2) is

$$x(t) = e^{-dt}\left(A \sin \omega_d t + B \cos \omega_d t\right),$$

where A and B are constants and

$$\omega_d = \sqrt{\omega^2 - d^2}. \tag{21.3}$$

The period and frequency of the damped vibrations are

$$\tau_d = \frac{2\pi}{\omega_d} \quad \text{and} \quad f_d = \frac{\omega_d}{2\pi}.$$

It is clear from Eq. (21.3) that $\omega_d < \omega$, so that the *period of the vibration is increased and its frequency is decreased* as a result of subcritical damping.

— **Supercritical Damping.** If $d > \omega$, the system is said to be *supercritically damped*. The general solution of Eq. (21.2) is

$$x(t) = Ce^{-(d-h)t} + De^{-(d+h)t},$$

where C and D are constants and

$$h = \sqrt{d^2 - \omega^2}.$$

— **Critical Damping.** If $d = \omega$, the system is said to be critically damped. The general solution is

$$x(t) = Ce^{-dt} + Dte^{-dt},$$

where C and D are constants.

Note. The *rate of damping* is often expressed in terms of the logarithmic decrement δ, which is the natural logarithm of the ratio of the amplitude at a time t to the amplitude at time $t + \tau_d$

$$\delta = \ln\left[\frac{e^{-dt}}{e^{-d(t+\tau_d)}}\right] = d\tau_d.$$

21.3 Forced Vibrations

The term *forced vibrations* means that external forces affect the vibrations of a system. Up until now, this chapter has discussed only *free vibrations* of systems: vibrations unaffected (*free*) from external forces.

- The *forced vibrations* of many damped one-degree-of-freedom systems are governed by the equation

$$\frac{d^2x}{dt^2} + 2d\frac{dx}{dt} + \omega^2 x = a(t), \qquad (21.4)$$

  where $a(t)$ is the forcing function and d and ω are given.

  — The general solution of Eq. (21.4) is the sum of homogeneous and particular solutions

$$x(t) = x_h(t) + x_p(t).$$

  The homogeneous solution x_h is the general solution of Eq. (21.4) when the right-hand side is set to zero. A particular solution $x_p(t)$ of Eq. (21.4) is any solution that satisfies Eq. (21.4).

Oscillatory Forcing Function

Here we consider the motion of a vibrating system subjected to an oscillatory forcing function and determine the response of the system as a function of the frequency of the force.

- **Particular Solution**. If, in Eq. (21.4), $a(t)$ is an oscillatory function of the form

$$a(t) = a_0 \sin\omega_0 t + b_0 \cos\omega_0 t,$$

where a_0, b_0, and ω_0 (the frequency of the forcing function) are given constants, the particular solution is

$$x_p(t) = \left[\frac{\left(\omega^2 - \omega_0^2\right) a_0 + 2d\omega_0 b_0}{\left(\omega^2 - \omega_0^2\right)^2 + 4d^2\omega_0^2}\right] \sin \omega_0 t + \left[\frac{\left(\omega^2 - \omega_0^2\right) b_0 - 2d\omega_0 a_0}{\left(\omega^2 - \omega_0^2\right)^2 + 4d^2\omega_0^2}\right] \cos \omega_0 t,$$

and its amplitude is

$$E_p = \frac{\sqrt{a_0^2 + b_0^2}}{\sqrt{\left(\omega^2 - \omega_0^2\right)^2 + 4d^2\omega_0^2}}$$

Note 1. Resonant Frequency. The frequency at which the amplitude of the particular solution is a maximum is called the *resonant frequency.*

Note 2. The particular solution for the motion of a damped vibrating system subjected to an oscillatory external force (forcing function) is also called the *steady-state solution.*

The motion approaches the steady-state solution with increasing time.

Polynomial Forcing Function

- **Particular Solution**. If, in Eq. (21.4), $a(t)$ is a polynomial forcing function of the form

$$a(t) = a_0 + a_1 t + a_2 t^2 + \cdots + a_N t^N,$$

where $a_1, a_1, \ldots, a_N$ are constants, the particular solution can be obtained by seeking a solution of the same form:

$$x_p(t) = A_0 + A_1 t + A_2 t^2 + \cdots + A_N t^N,$$

where $A_0, A_1, \ldots, A_N$ are constants that must be determined.

Helpful Tips and Suggestions

- When dealing with (small) undamped free vibration, the differential equation describing vibrations always takes the standard form Eq. (21.1). This provides a useful way of 'checking' your work when modelling and analyzing vibrations of a similar vibrating system.
- It may be useful to review some elementary techniques for solving linear ordinary differential equations with constant coefficients.

Review Questions

1. What is a one-degree-of-freedom system?
2. When does free vibration occur?
3. Write down the differential equation describing undamped free vibrations, its general solution and the definitions of "amplitude," "period," and "frequency".
4. What is simple harmonic motion?
5. Why is it possible to use conservation of energy to obtain the (natural) frequency of a body executing undamped free vibrations?
6. What is the resonant frequency ?
7. How do you determine whether the damping of a system whose motion is described by Eq. (21.2) is subcritical, critical, or supercritical?
8. What is the logarithmic decrement?
9. Describe how the general solution of Eq. (21.4) splits into parts.
10. If a vibrating system is subjected to a polynomial forcing function. how do you determine the particular solution of the equation describing the system's motion i.e., Eq. (21.4).

Answers to Review Questions

- Chapter 12

  1. F 2. T 3. T 4. F 5. T 6. F 7. F 8. T 9. F 10. T

- Chapter 13

1. Yes because a particle is modelled as a point (i.e., an object with mass but negligible size/shape).
2. Yes.
3. No. For example, a particle travelling in a general curvilinear path has both normal and tangential acceleration components.
4. No. Since the ball has constant speed v and, from Section 13.3 $a_t = \frac{dv}{dt}$ and $a_n = \frac{v^2}{\rho}$, the tangential component of acceleration is always zero **but** the normal component of acceleration is non-zero. This makes the acceleration of the center of the ball non-zero.
5. No. The velocity depends on the reference frame. For example, we can always define a reference frame with respect to which the particle isn't moving (i.e., the particle stays fixed for all time in that reference frame), in which case, the particle's velocity (and acceleration) will always be zero.
6. This happens when $a_n = \frac{v^2}{\rho} \equiv 0 \Longleftrightarrow v = 0$ or $\rho \to \infty$ i.e., when the particle is fixed for all time or when the radius of curvature is infinite i.e., when the particle is moving in a straight line (rectilinear motion).
7. No. For example, if the rectilinear motion of a point P is described by $x\,(t) = t^2$, then

$$\text{speed: } \dot{x}\,(t) = 2t$$
$$\text{acceleration: } \ddot{x}\,(t) = 2$$

   Clearly the speed is zero at $t = 0$ but the acceleration is constant ($= 2$) for all time.
8. From Eq. (12.8), the radial component of acceleration for general curvilinear motion is given by $a_r = \ddot{r} - r\dot{\theta}^2$. If $\ddot{r} = 0$, $a_r = -r\dot{\theta}^2 \neq 0$.
9. No—see Question 4.
10. Angular velocity measures the angular motion of a line whereas the velocity of a point measures the rate of change of position of the point.

- Chapter 14

1. False. The second law is based *solely* on experimental evidence.
2. Center of mass.
3. The first two equations of Eqs. (14.4).
4. When an object moves over a *known planar curved path.*
5. Observation (over twenty years!).

- Chapter 15

1. See Section 15.2. i.e., $U_{1-2} = \int_{\mathbf{r}_1}^{\mathbf{r}_2} \sum \mathbf{F} \cdot \mathbf{dr} = \int_{s_1}^{s_2} \sum F_t ds$, where $\sum F_t$ is the tangential component of the total external force on the object.
2. The principle of work and energy is used to solve kinetic problems that involve *velocity, force, and displacement* (since these terms are involved in the equation describing the principle—see Section 15.1).
3. No work is done by the forces perpendicular to path, so the principle of work and energy (Eq. (15.2) states that the kinetic energy remains constant.
4. *Power* is the rate at which work is done. Hence, the power P generated by a machine or engine which performs an amount of work dU within a time interval dt is given by

$$P = \sum \mathbf{F} \cdot \mathbf{v}$$

   where $\mathbf{v}$ is the velocity of the point which is acted upon by the system of external forces $\sum \mathbf{F}$.
5. When the work done by a force in moving an object from one point to another is *independent of the path* followed by the object, then this force is called a *conservative force.* e.g., weight of an object or spring force acting on an object (see Section 15.5).
6. The work done by the *weight of an object* is *independent of the path* of the object i.e., the work done depends only on the object's *vertical displacement.*
7. The *force of friction* exerted *on a moving object* by a fixed surface *depends on the path* of the object i.e., the longer the path, the greater the work. Consequently, frictional forces are *nonconservative*. The work is dissipated from the body in the form of heat.
8. *Potential energy* is a reservoir of "potential" kinetic energy e.g., gravitational potential energy, elastic potential energy (see Section 15.4).
9. A force $\mathbf{F}$ is conservative if and only if $curl\mathbf{F} = \mathbf{0}$ (see Section 15.6).
10. The conservation of energy equation is used to solve problems involving velocity, displacement and *conservative force systems*—all of which form part of the energy equation (15.3).

- Chapter 16

1. A particle's linear momentum is described by the *vector* $\mathbf{L} = m\mathbf{v}$. It's magnitude is mv and its direction is the same as that of the velocity $\mathbf{v}$.
2. The *principle of linear impulse and momentum* is used to solve problems involving *force, time and velocity.* It provides a *direct means* of obtaining the particle's final velocity $\mathbf{v}_2$ after a specified time period when the particle's initial velocity is known and the forces acting on the particle are either constant or can be expressed *as functions of time.*
3. See end of Section 16.1.
4. When the sum of the external impulses acting on a system of particles is zero or negligible.
5. The velocity of the common center of mass is constant.

6. If colliding objects A and B are not subjected to external forces, *their total linear momentum must be the same before and after the impact* (total linear momentum of the system composed of objects A and B is *conserved*). Even when the colliding objects A and B are subjected to external forces, the force of the impact is often so large, and its duration so brief, that the effect of the external forces on their motions during the impact is *negligible*. Hence, the velocity of their common center of mass before and after the impact is given by Eq. (16.3) i.e.

$$\mathbf{v} = \frac{m_A \mathbf{v}_A + m_B \mathbf{v}_B}{m_A + m_B} = \text{constant.}$$

7. The angular momentum $\mathbf{H}_O$ of an object about point O is defined as the *moment* of the particle's linear momentum about O. It is sometimes referred to as the moment of momentum.
8. From Eq. (16.4), when the angular impulses acting on an object are all zero during the time t_1 to t_2.
9. False. For example, when the object is subjected only to a central force. Then the impulsive central force **F** is always directed towards point O as the particle moves along the path. Hence, the angular impulse (moment) created by **F** about the *z axis* passing through point O is always zero and therefore angular momentum of the particle is always conserved about the *z axis*, but linear momentum is not.
10. It's constant (conserved).

- Chapter 17

1. A rigid body is an idealized model of an object that does not deform or change shape.
2. If the points of a rigid body intersected by a fixed plane remain in that plane, the rigid body is said to undergo two-dimensional or *planar* motion.
3. The angular velocity vector ω of a rigid body is parallel to the *instantaneous axis of rotation of the body*, and its magnitude $|\omega|$ is the body's *rate of rotation.*
4. If the thumb of the right hand points in the direction of ω, the fingers curl around ω in the direction of the rotation.
5. Use Equation (17.1).
6. Zero.
7. No—it may be "at infinity" (corresponding to 'translation' of the rigid body at that instant).
8. No—since the instantaneous center does not, in general, have zero acceleration.
9. The velocity of A relative to the body-fixed reference frame is zero i.e., A moves only with the rigid body and not relative to it.
10. The angular acceleration vector $\alpha = \dfrac{d\omega}{dt}$ is the *rate of change of the angular velocity vector* of the body.
11. A reference frame attached to the rigid body (and therefore moves with the rigid body).
12. The acceleration of A relative to the secondary reference frame (e.g., body-fixed reference frame)—one which moves relative to some primary reference frame.

- Chapter 18

1. Since any object or collection of objects, including a rigid body, can be modeled as a system of particles.
2. Eqs. (18.3) and (18.5).
3. 1. **Draw the Free-Body Diagram**. Isolate the object and identify the external forces and couples acting on it.
   2. **Apply the Equations of Motion.** Write equations of motion suitable for the type of motion, choosing an appropriate coordinate system for applying Newton's second law. For example, if the center of mass moves in a circular path, it may be advantageous to use normal and tangential components.
   3. **Determine Kinematic Relationships.** If necessary, supplement the equations of motion with relationships between the acceleration of the center of mass and the angular acceleration of the object.

4. Eq. (18.3) (Newton's second law and possibly $\sum M = 0$.
5. Only Eq. (18.4) and possibly Newton's second law.
6. If the rigid body rotates about O.
7. Three (two of force and one of angular motion).
8. Six (three of force and three of angular motion).

- Chapter 19

1. The work done by external forces and couples as a rigid body moves between two positions equals the change in the total kinetic energy of the body.
2. Eq. (19.1).
3. Constant.
4. Conserved or constant.
5. There may be significant external forces acting that produce linear impulses e.g., if one of the rigid bodies has a pin-support, the reactions exerted by the pin support cannot be neglected and linear momentum is not conserved.
6. No—see answer to Question 5.
7. Yes—since the reactions at O contribute no external angular impulse *about* O.
8. Their velocities at the point of impact.

- Chapter 20

1. $\omega = \Omega + \omega_{\text{rel}}$.
2. The three-dimensional equations of motion for a rigid body.
3. See the inertia matrix in Section 20.2.
4. Eqs. (20.1) are for a rigid body *rotating about a fixed point* O, whereas Eqs (20.2) are for general three-dimensional motion.
5. All vectors are three component vectors in three-dimensional kinematics, as opposed to two component vectors in plane kinematics.
6. Euler angles are used to specify the orientation of a rigid body in three-dimensions.
7. In the *steady precession* of an axisymmetric spinning object, the spin rate $\dot{\phi}$, the nutation angle θ and the precession rate $\dot{\psi}$ are *assumed to be constant*.

- Chapter 21

1. If a vibrating system has *one degree of freedom*, it requires only *a single coordinate* (x) to specify completely the position of the system at any time t.
2. Free vibration occurs when the motion is maintained by gravitational or elastic restoring forces, that is, the vibration is *free* from external periodic or intermittent forces applied to the system.
3. The differential equation describing undamped free vibrations is $\ddot{x} + \omega^2 x = 0$. Its general solution and other vibrational characteristics are

$$x(t) = A \sin \omega t + B \cos \omega t,$$

$$x(t) = E \sin(\omega t - \phi),$$

$$E = \sqrt{A^2 + B^2} \text{ is the } \textit{amplitude},$$

$$\tau = \frac{2\pi}{\omega} \text{ is the } \textit{period},$$

$$f = \frac{1}{\tau} \text{ is the } \textit{(natural) frequency}.$$

4. Simple harmonic motion describes a situation where acceleration is proportional to displacement.
5. Since in undamped free vibrations the motion is due to *only* gravitational and elastic restoring forces, which are conservative.
6. The frequency at which the amplitude of the particular solution is a maximum.
7. Subcritical: $d < \omega$, supercritical: $d > \omega$, critical: $d = \omega$.
8. The *rate of damping* is often expressed in terms of the logarithmic decrement δ, which is the natural logarithm of the ratio of the amplitude at a time t to the amplitude at time $t + \tau_d$:

$$\delta = \ln\left[\frac{e^{-dt}}{e^{-d(t+\tau_d)}}\right] = d\tau_d.$$

9. The general solution of Eq. (21.4) is the sum of homogeneous and particular solutions:

$$x(t) = x_h(t) + x_p(t).$$

The homogeneous solution x_h is the general solution of Eq. (21.4) when the right-hand side is set to zero. A particular solution $x_p(t)$ of Eq. (21.4) is any solution that satisfies Eq. (21.4).

10. By seeking a solution of the same form:

$$x_p(t) = A_0 + A_1 t + A_2 t^2 + \cdots + A_N t^N,$$

where $A_0, A_1, \ldots, A_N$ are constants that must be determined.